The Home Plumbing Handbook

Charles N. McConnell

DELMAR
THOMSON LEARNING™

Australia Canada Mexico Singapore Spain United Kingdom United States

DELMAR
THOMSON LEARNING™

The Home Plumbing Handbook

Charles N. McConnell

Vice President, Technology and Trades SBU:
Alar Elken

Editorial Director:
Sandy Clark

Development:
Jennifer Luck

Marketing Director:
Cyndi Eichelman

Marketing Coordinator:
Brian McGrath

Production Director:
Mary Ellen Black

Production Manager:
Andrew Crouth

Production Editor:
Stacy Masucci

Printed in Canada
2 3 4 5 6 XX 06 05 04 03

For more information contact
Delmar Learning
Executive Woods
5 Maxwell Drive, PO Box 8007,
Clifton Park, NY 12065-8007
Or find us on the World Wide Web at
www.delmarlearning.com

Library of Congress Cataloging-in-Publication Data:

McConnell, Charles.
Home plumbing handbook / Charles N. McConnell.
p. cm.
Includes index.
ISBN 1-4018-5625-X
1. Plumbing—Amateurs' manuals. I. Title.

TH6124.M3 2004
696'.1—dc21

2003052058

NOTICE TO THE READER

Publisher does not warrant or guarantee any of the products described herein or perform any independent analysis in connection with any of the product information contained herein. Publisher does not assume, and expressly disclaims, any obligation to obtain and include information other than that provided to it by the manufacturer.

The reader is expressly warned to consider and adopt all safety precautions that might be indicated by the activities herein and to avoid all potential hazards. By following the instructions contained herein, the reader willingly assumes all risks in connection with such instructions.

The publisher makes no representation or warranties of any kind, including but not limited to, the warranties of fitness for particular purpose or merchantability, nor are any such representations implied with respect to the material set forth herein, and the publisher takes no responsibility with respect to such material. The publisher shall not be liable for any special, consequential, or exemplary damages resulting, in whole or part, from the readers' use of, or reliance upon, this material.

Contents

Preface

This book is different from other plumbing books because the author is a plumber who actually worked at the trade for 40 years, both as a journeyman and master plumber. When you start a home improvement project, things often don't work out just the way you expected. One job frequently leads to another. As the saying goes: Your author has been there and done that many, many times. He has encountered the pitfalls that await you as you start a project, and he will be there to help you overcome them.

Turning off the water to a faucet or a fixture is usually the first step in a plumbing home improvement project. Because it may also be the hardest part of that project, I have included a chapter showing how to solve this problem.

I have written this book for the do-it-yourselfer. In this do-it-yourself age virtually everything needed in home construction or remodeling, from the materials to the tools needed, can be found in a nearby home improvement store and 90 percent of the materials purchased there are installed by the buyer.

Fully one-half the cost of making needed or desirable improvements is labor cost. You can make the desired improvements yourself and save hundreds of dollars in the process; the secret is having concise, clear, easy-to-follow instructions. As a licensed journeyman and master plumber

I have done every project in this book hundreds of times, and I know the pitfalls that await the novice. I can tell you ahead of time the things you are likely to run into, and this knowledge will help you make that plumbing improvement a pleasant experience instead of a nightmare.

The Appendix at the back of the book can help solve many problems. If you do not know what kind of faucet you have, there are life-size drawings of faucet stems to help you identify it. You will find faucet stems, seats, gaskets, O rings, and virtually every part to repair any faucet made in the last 50 years listed in the Appendix. And any good home improvement store will have these parts in stock.

PLEASE NOTE: Some of the operations covered in this book require experience and training equal to that of a licensed tradesman in the field. Disconnection and reconnection of electrical wiring or the installation of new wiring may be required. Before starting on any project where electrical wiring is involved, *remove* the fuse or turn the circuit breaker controlling the circuit involved to OFF position. The installation and repairs explained in this book must comply with local codes and regulations.

About the Author

Charles N. McConnell has held Journeyman and Master Plumbers licenses in Indiana and a Master Plumbers license in Florida. He is the recipient of an award for over 60 years of continuous membership in the United Association of Journeymen and Apprentices of the Plumbing and Pipefitting Industry of the United States and Canada.

His experience qualifies him as an expert witness in litigation of plumbing-related court cases. He has designed, supervised, and installed plumbing, heating, and air conditioning in residential, commercial, industrial, and public buildings and has trained many apprentices for the plumbing trade.

Other books by this author:

Plumbers and Pipefitters Library, 3 vols. (Macmillan)

Pipe Fitters and Welders Pocket Manual (Macmillan)

Plumbers Maintenance/Troubleshooting Pocket Manual (Macmillan)

Building an Addition to Your Home (Prentice-Hall)

Acknowledgments

The author wishes to thank the following companies for their assistance in furnishing information, drawings, and photos on their products.

Anaheim Manufacturing Co.
P.O. Box 4146
Anaheim, CA 92803
(800) 854-3229

A.O. Smith
11270 West Park Place
Milwaukee, WI 53224
(414) 359-4000
Fax (414) 359-4064

Delta Faucet Co.
P.O. Box 40980
Indianapolis, IN 46280
www.deltafaucet.com

Fluidmaster, Inc.
30800 Rancho Viejo Road
San Juan Capistrano, CA 92675

Goulds Pumps/ITT Industries
240 Fall Street
Seneca Falls, NY 13148

In Sink Erator
4700 21st St.
Racine, WI 53406

JET® INC
750 Alpha Drive
Cleveland, OH 44143

Mueller Industries, Inc.
8285 Tournament Drive, Suite 150
Memphis, TN 38125
(901) 753-3200

Peerless Pottery
P.O. Box 145
Rockport, IN 47635-0145

Price Pfister
19701 DaVinci
Lake Forest, CA 92610
(949) 672-4000
(800) Pfaucet
Fax (800) 713-7080

Radiator Specialty Co.
1900 Wilkinson Boulevard
Charlotte, NC 28234-4689

Sterling Plumbing Group, Inc.
2900 Gulf Rd.
Rolling Meadows, IL 60008

Watts Regulator
815 Chestnut Street
North Andover, MA 01845-6098

1

Basic Plumbing Information

The plumbing system is the most important part of our homes; it provides us with drinking and cooking water and gets rid of our water-carried wastes. We take our plumbing systems for granted until a stopped-up toilet, a backed-up kitchen sink, or a faucet that leaks like Niagara Falls creates havoc.

Hundreds of thousands of us have discovered that we can handle these emergencies and make needed repairs, install new appliances, and even tackle a bathroom and/or kitchen remodeling project *if we have good instructions.* And we can do the job as well as the professionals would.

To help you understand the basics of plumbing systems, the best place to start is with the drainage and vent systems in a typical home. A basement and single story home is shown in Figure 1–1. Your home may be larger, but the principles are still the same. I use plumber's terminology throughout this book. I've stood at parts counters in supply houses and hardware stores listening to someone trying to describe "this gizmo that screws into that gadget"—very frustrating for both the clerk and the customer. When you need parts or supplies, *know what to ask for.*

The Basic System Parts

The drainage system includes all the piping within public or private premises, which conveys sewage or other liquid wastes to a legal point of disposal but does not include the mains of a public sewer system or a public sewage treatment or disposal plant. The *building drain* is that part of the lowest piping of a drainage system which receives the discharge from soil, waste, and other drainage pipes inside the walls of the building and conveys it to the building sewer beginning 2 feet (610 mm) outside the building wall.

A *soil pipe* is any pipe which conveys the discharge of water closets, urinals, clinic sinks, or fixtures having similar functions of collection and removal of domestic sewage, with or without the discharge from other fixtures to the building drain or building sewer.

A *waste pipe* is a pipe that conveys only liquid waste that is free of fecal matter.

A *stack* is a vertical extension of a soil, waste, or vent pipe.

A *vent pipe* is any pipe provided to ventilate a plumbing system, to prevent trap siphonage and back pressure, or to equalize the air pressure within the drainage system.

A *trap seal* is the maximum vertical depth of liquid that a trap will retain, measured between the crown weir and the top of the dip of the trap.

On the left side of Figure 1–1, just inside the building wall, a cleanout fitting extends up to floor level from A, the building drain. On the right in Figure 1–1, a soil and waste stack, B, receives the discharge from the bathroom fixtures. You will notice that there is a cleanout at the base of the stack. Plumbing codes require a cleanout at the base of every stack. The soil stack receives the discharge from the toilet and the bathtub. The lavatory waste empties into the vent; thus the vent serving the toilet and tub becomes a *wet* vent. Both A and B extend up and out through the roof. A roof flashing, H, prevents water leaks at this point. The flashing shown is called a *boot* flashing; it is made of lead, and the top is turned down into the vent pipe. In areas where freezing temperatures occur, if the vent pipe below the roof is 2 inches or smaller, a fitting called an increaser, G, is used to extend the vent pipe through the roof. The increaser is used to prevent frost from forming and closing the vent pipe.

Why are traps installed on plumbing fixtures? Contrary to what you might think, traps are not there to "catch" anything. Traps prevent sewer gas from entering a building through drain piping.

You may wonder why the waste pipe C cannot serve as a wet vent for the washer drain. Water released from the sink could drop down the stack, acting like a plug, and fill the pipe, pushing air ahead of it and creating a vacuum which could cause back-siphonage and pull the water from the trap serving the washer. With no water in the trap, sewer gas could then enter the building through the trap. Proper venting of fixtures protects health and provides safety. (See Chapter 9, "Protecting Your Family's Health.")

The piping sizes shown in Figure 1–1 are intended to serve as a guide and in general represent minimum acceptable sizes.

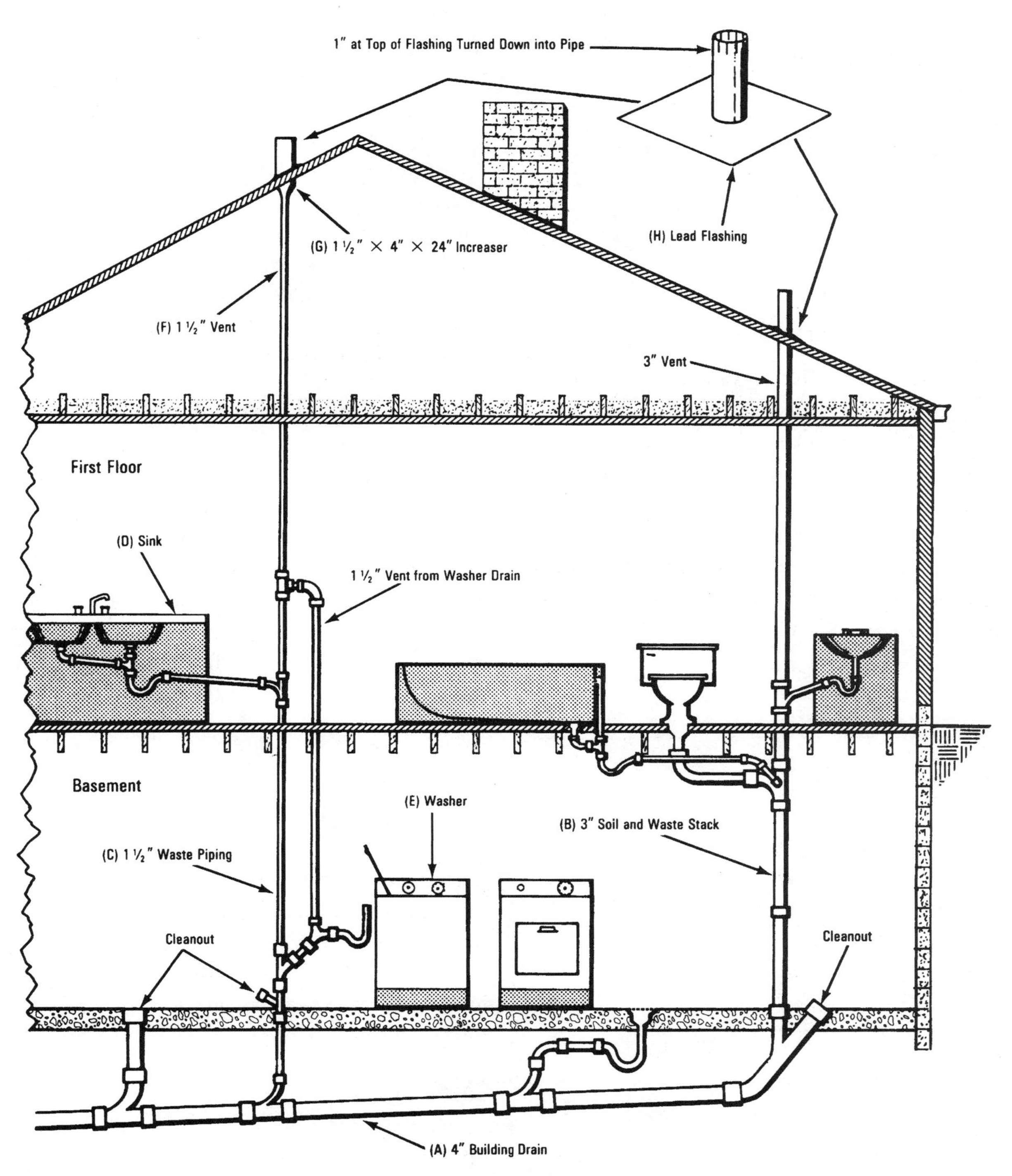

Figure 1–1 Drainage and vent system in a basement and single story home.

You should always check local codes for specific requirements in your area.

Plumbing Valves and Their Uses

As seen in Figure 1–2, valves in plumbing systems have four specific functions:

1. Flow is turned ON.
2. Flow is throttled or regulated.
3. Flow is turned OFF.
4. Flow is permitted in one direction only.

The Main Water Valve

The water service pipe is the cold water line from a water main in the street, a well, or other source, into a building. A control valve, which will control all the water lines in the building, should be installed at or near the entrance point of the service line into the building. The type of valve specified in this location will depend on the water utility or the local code. This valve may be a "full port" valve (when valve is open, there should be no restriction to flow) or it may be a globe valve. Learn where this valve is and mark or tag it so that it can be easily and quickly located in the event of an emergency. The cold water piping will continue from this point to the fixtures requiring cold water connections.

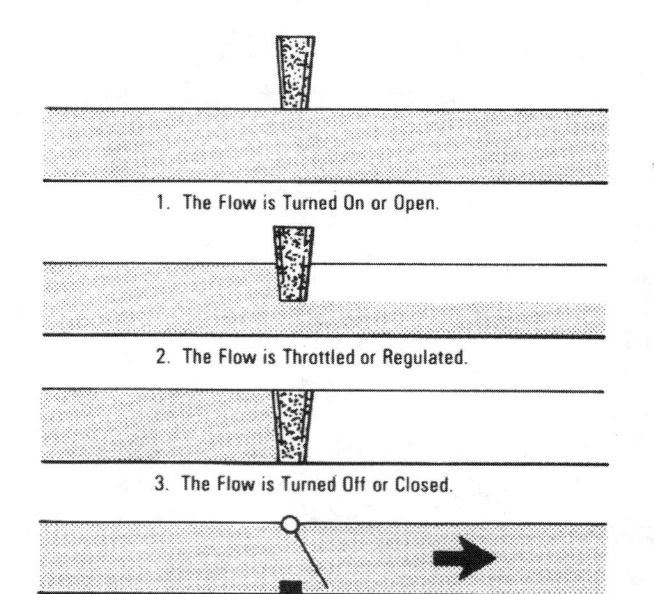

Figure 1–2 Four functions of valves.

Hot-Water Control Valve

The hot-water control valve, located on the *cold water supply* to the water heater, controls all the hot water to the house. In the event of a leak in the water heater, a broken or leaking hot water pipe, or the need to shut off hot water to a fixture while making repairs, closing this valve will shut off all the hot water to the building, but will leave the cold water supply to the building on.

Gate Valves

One of the many types of gate valves is shown in Figure 1–3. Gate valves are designed for full ON or OFF use; they should *not* be used as throttling valves. Gate valves may not shut off completely; the slight leak that can occur can cause a "wire cut" in the seat face, which compounds the chance of a leak. Gate valves are rarely used in residential plumbing.

Check Valves

Check valves are designed to permit flow in one direction only. One type of check valve is seen in Figure 1–4.

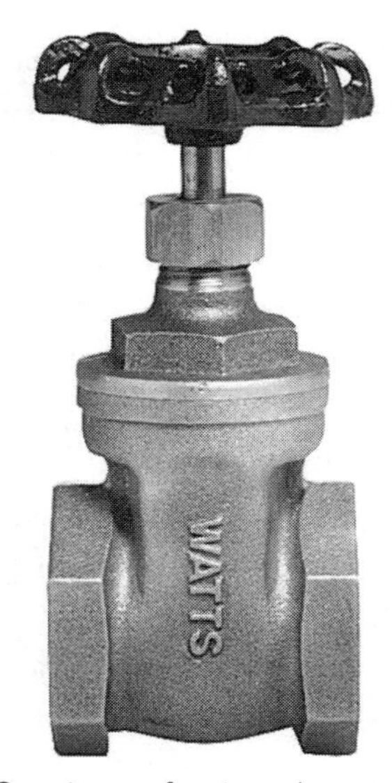

Figure 1–3 One type of gate valve.
Courtesy Watts Regulator

Figure 1–4 One type of check valve.
Courtesy Watts Regulator

Ball Valves

Ball valves are a relatively new and valuable addition to the valve family and are available in both metal and plastic. Ball valves can be used as throttling valves or in full OFF or ON position. *Note:* All ball valves are not full port valves. In full ON position there is no restriction to flow using a full port valve. The polyvinyl chloride (PVC) ball valve shown in Figure 1–5 has sockets on each end, and the valve in Figure 1–6 is a union type with coupling nuts on each end of the valve.

Figure 1–5 This type of PVC ball valve has a socket on each end.
Courtesy Watts Regulator

Relief Valves

Probably the most important valve in the home is the temperature and pressure relief valve, Figure 1–7, installed on your hot water heater. There are many instances on record where a home or a building was destroyed when the wrong relief valve was installed. Relief valves are used on water

Figure 1–6 This is a union-type PVC ball valve.
Courtesy Watts Regulator

Figure 1–7 A temperature and pressure relief valve.
Courtesy Watts Regulator

heaters and pressure tanks as safety valves. If the thermostat on a water heater or a pressure switch on a well system should malfunction, excessive pressure could build up, causing a water heater or pressure tank on a well system to explode. Relief valves in these systems are designed to open and relieve excess pressure.

Globe Valves and Compression Stops

Globe valves, and variations of globe valves called *compression stops* (see Figure 1–8), are widely used in residential plumbing. Although there is some flow restriction when globe valves or compression stops are used, the restriction is so slight that it is hardly noticeable. Globe valves and compression stops are generally installed for full ON or OFF use; however, globe valves can be used for throttling purposes. Globe valves should be installed with the stem vertical and with the correct direction of flow as shown by an arrow on the valve. One type of globe valve, a compression stop, is commonly used as a control valve on bath and shower units. This valve is shown in Figure 1–8.

Two types of globe valves or compression stops are shown in Figure 1–9. These types are commonly used to control sink, lavatory, and tank-type toilet supplies and other fixtures using 1/2- or 3/8-inch flexible supplies.

Problems with Compression Stops

When stops of the types shown in Figure 1–9 have not been turned OFF or ON for a long period of time, they often become corroded, making it very difficult to turn the handles. Caution is necessary here: If the stop is soldered to copper tubing, excessive

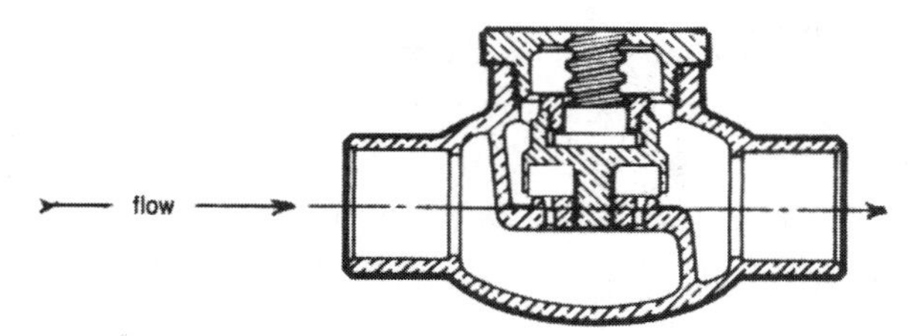

Figure 1–8 A globe valve.
Courtesy Nibco, Inc.

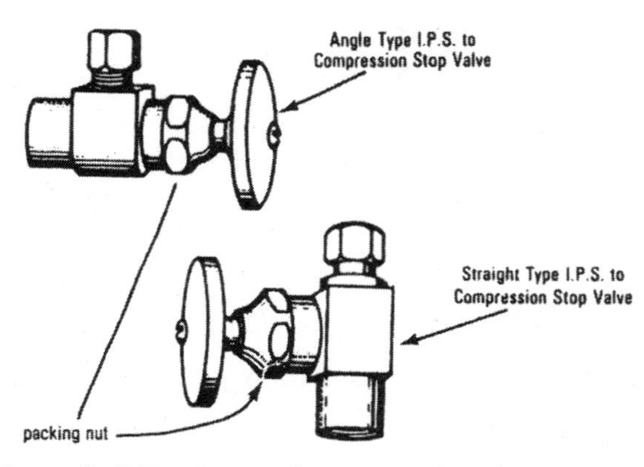

Figure 1–9 Two types of compression stops.

force could twist the copper tubing like a corkscrew, making replacement of the stop necessary. If the stop is screwed onto a threaded nipple and the stop turns, it could cause a leak. Because this is a very common problem, I have devoted an entire chapter, Chapter 3, to the problem.

Sillcocks

There are two types of sillcocks for hose connections. The angle sillcock, Figure 1–10, is not freezeproof; *garden hoses should always be disconnected from this type sillcock when freezing temperatures are expected.*

The freezeproof sillcock, which is 12 inches long, Figure 1–11, is designed to be used with the shutoff point in a warm area. When this type of sillcock is used and a garden hose is attached, *the hose must be removed during freezing weather to allow water in the sillcock to drain.* If a hose is left connected, the water in the sillcock will freeze and burst the sillcock.

Figure 1–10 A copper sweat-type angle sillcock.
Courtesy Watts Regulator

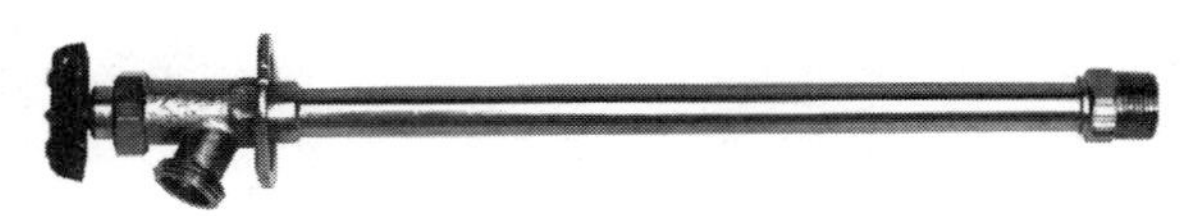

Figure 1–11 An antifreeze sillcock.
Courtesy Watts Regulator

Stop and Waste Valves

There are two types of stop and waste valves: automatic and button. Stop and waste valves are designed to turn off the water downstream of the valve and allow water in the pipe to drain. When either type of valve is turned off, pressure remains in the pipe until it is released. With an *automatic* stop and waste, when the valve is closed, water will start to drain, releasing the pressure. As the pressure is released and water drips from the pipe, a vacuum will form in the pipe. If this vacuum is not broken by opening a faucet, sillcock, or so on, water will not drain from the pipe.

When a button-type stop and waste valve, shown in Figure 1–12, is turned off, pressure will remain in the pipe until it is released. When the button, a threaded cap, is turned counterclockwise, the pressure will be released and water will drip from the pipe, forming a vacuum in the pipe. If a faucet or sillcock downstream of the valve is not opened to break the vacuum, water will not drain from the pipe. Arrows on the side of stop and waste valves indicate the direction of flow.

NOTE: Stop and waste valves are a potential source of backflow if installed below ground level. See Chapter 9, "Protecting Your Family's Health."

Figure 1–12 A button-type stop and waste valve.
Courtesy Watts Regulator

2

Tools to Make the Job Easy

Not too many years ago when you wanted to make repairs or installations in your home plumbing system, you needed a truck full of tools such as pipe dies, pipe vises, and heavy awkward pipe wrenches to make the plumbing repairs and installations we did then. But, times and materials have changed. Lightweight PVC pipe and fittings have taken the place of heavy steel pipe and fittings. We seldom use threaded pipe and fittings today; instead we use copper tubing and solder fittings, PVC, chlorinated PVC (CPVC), and PVC drainage-waste-vent (PVC-DWV) pipe and fittings that are glued together.

Today, the average do-it-yourselfer has a workbench with drawers or boxes containing the tools he or she has acquired over time, tools such as blade and Phillips-type screwdrivers, and a claw hammer, that are needed for household repairs other than plumbing. Almost every tool collection today contains a battery-type drill that doubles as a screwdriver.

Some special tools are needed when doing plumbing work. I will help you select those tools that will make your work of plumbing repair and installation easier and help you do a better job.

The following tools are those that will be needed at one time or another as you make needed repairs and

Figure 2–1 A 14-inch straight pattern pipe wrench and 14-inch end pipe wrench.
Courtesy Ridge Tool Co.

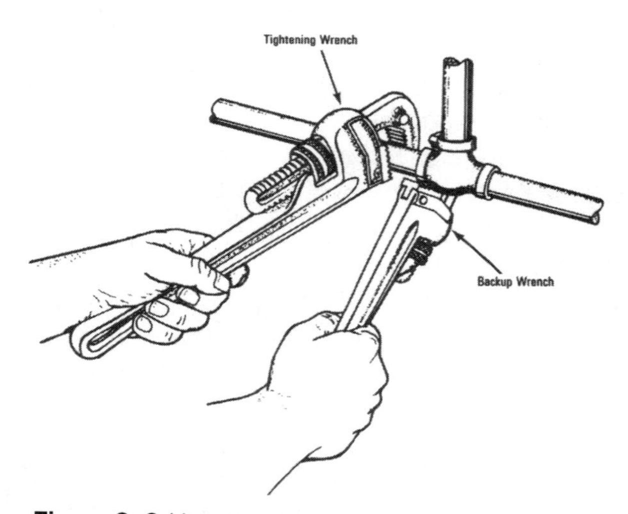

Figure 2–2 How to use a back-up wrench.

improvements. Don't let the list scare you; as I said earlier, you probably have some of these tools, and the ones you have have already paid for themselves. One thing is important: When you buy tools buy *good* tools.

If you become involved in a project that requires specialized tools (pipe threaders, pipe taps, vises, large pipe wrenches, etc.), consider renting them. Don't buy tools you may never need again.

The Tools You Should Have

Good pipe wrenches are a must. The wrenches shown in Figure 2–1 are 14-inch wrenches, but a 14-inch and a 10-inch wrench should be all you need around the house.

Two pipe wrenches are shown in Figure 2–2. The larger wrench, a 14-inch, is used to tighten or loosen pipe or fittings. The 10-inch back-up wrench is used to prevent a pipe or fitting that is already tight from being turned as another pipe or fitting is being tightened or loosened.

You must have slip-joint pliers in your tool box. (When you're working, they will be in your hip pocket.) See Figure 2–3.

You will also need a 6-foot inside-reading folding rule, Figure 2–4. It will lay flat on plans or on work, for accuracy in measuring.

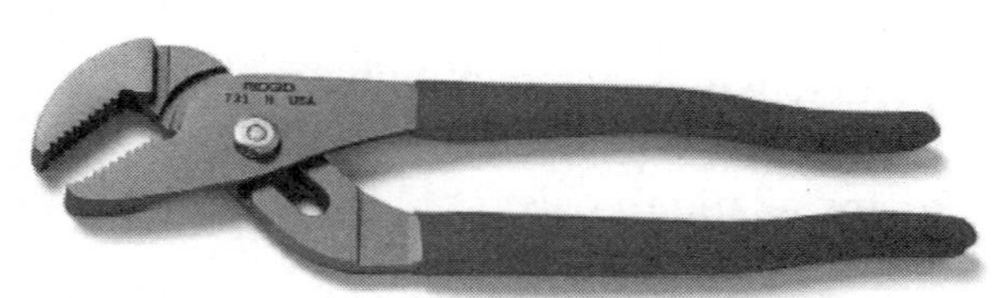

Figure 2–3 Slip-joint pliers.
Courtesy Ridge Tool Co.

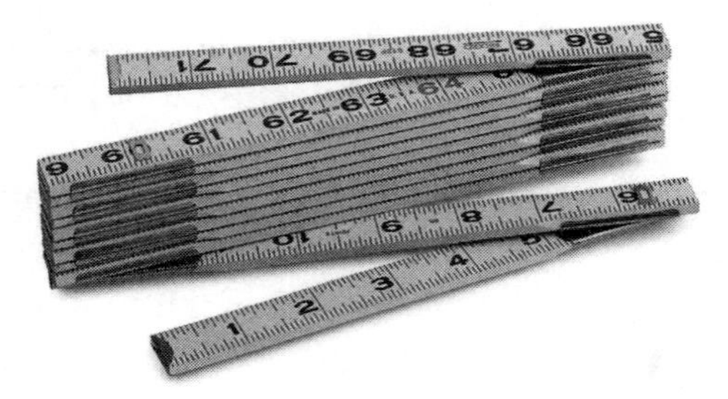

Figure 2–4 A 6-foot inside-reading rule.
Courtesy Ridge Tool Co.

The 20-foot locking-type steel measuring tape, Figure 2–5, is handy for measuring long lengths of pipe.

Because of the very tight working space in back of a kitchen sink or a bathroom lavatory, it is almost impossible to remove the faucets and install new ones without a basin wrench. The wrench shown in Figure 2–6 telescopes to make access to parts easier and has a spring-loaded jaw for grasping hard-to-get-at parts. No other tool takes the place of a good basin wrench.

Now there is a new basin wrench, Figure 2–7. It does not take the place of the one in Figure 2–6; it is a companion tool to it. The new basin wrench is designed to self-center on two, three, four, and six tab nuts and to fit metal hex nuts. It is only 11 inches long and fits neatly in a tool box tray. Once you have used this tool, you will never be without one.

The plastic pipe or tubing cutter shown in Figure 2–8 has a compound leverage ratchet mechanism and a hardened steel blade. One-hand operation produces quick, clean cuts with ease. One or two squeezes on the handle and the compound-leverage cutter cuts plastic pipe from 1/8 to 1 1/8 inch outside diameter (O.D.) quickly and easily.

The tubing cutter shown in Figure 2–9 can be used to cut copper, brass, aluminum, or plastic tubing and thin wall conduit. Two sizes cover normal household repair requirements.

The plungers shown in Figure 2–10 can exert tremendous hydraulic force to open stopped drains. If you are using them on a two-compartment sink, hold one plunger over an opening and exert force on the other one.

Figure 2–5 A locking-type steel measuring tape.
Courtesy Ridge Tool Co.

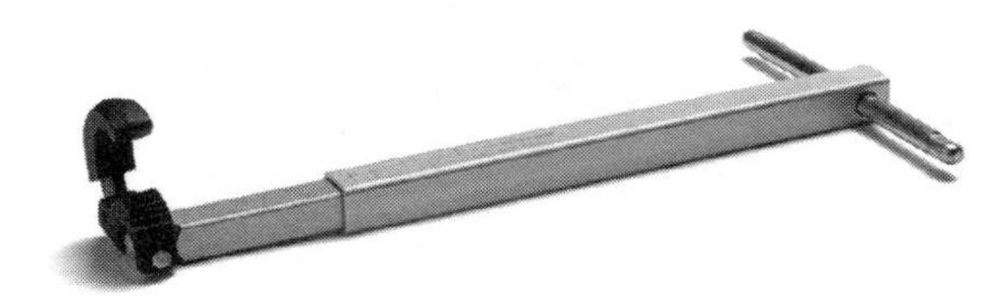

Figure 2–6 A telescoping basin wrench.
Courtesy Ridge Tool Co.

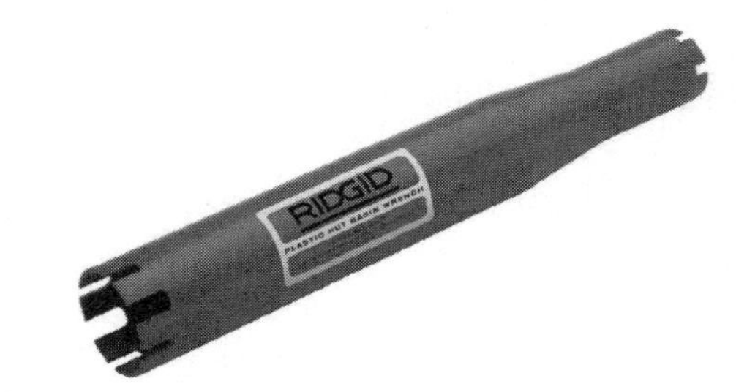

Figure 2–7 A new-type basin wrench.
Courtesy Ridge Tool Co.

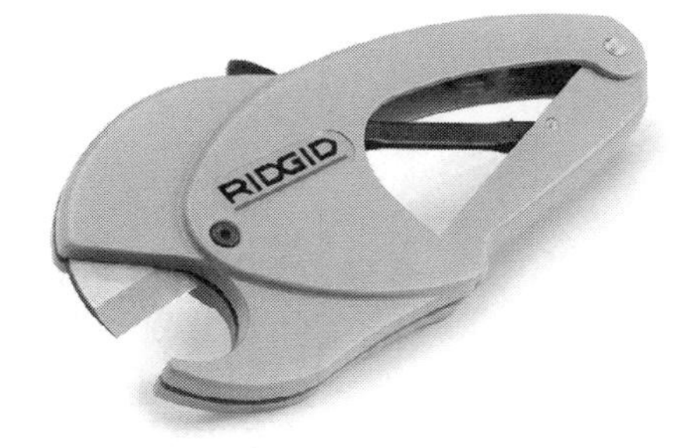

Figure 2–8 A compound lever plastic tubing cutter.
Courtesy Ridge Tool Co.

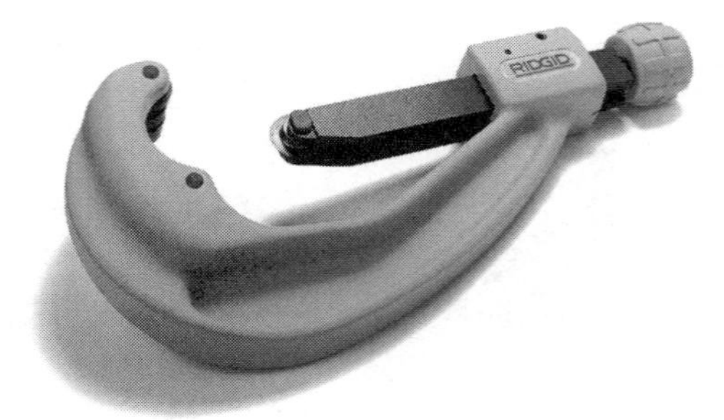

Figure 2–9 A copper and plastic tubing cutter.
Courtesy Ridge Tool Co.

Figure 2–10 Two plungers for unstopping drains.

Figure 2–11 A propane tank and burner.

The propane tank with a pencil burner, Figure 2–11, will apply heat just where it is needed for soldering copper tubing.

Adjustable wrenches, shown in Figure 2–12, are indispensable tools in plumbing work; they are used for taking faucets apart, tightening compression nuts, and so on.

We couldn't do plumbing work without a good hacksaw. The saw shown in Figure 2–13 has an adjustable frame to accommodate different-length blades. Coarse-toothed blades (18 teeth per inch) are used for cutting steel bars, pipe, or heavy metals. Medium-toothed blades (24 teeth per inch) are for pipe or metal up to 8 inches thick. Fine-toothed blades, (32 teeth per inch) are used for thin metals, thin tubing, and so on.

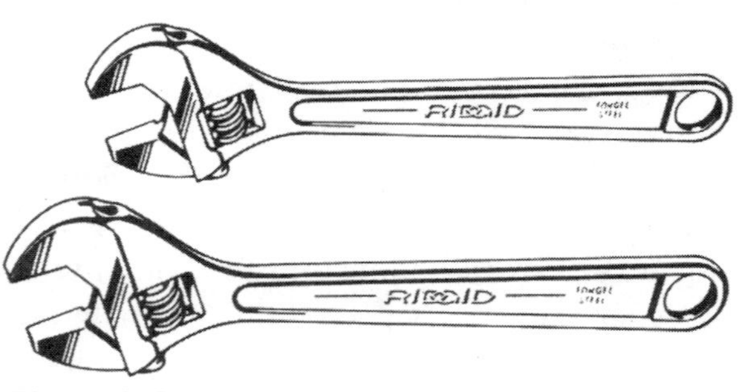

Figure 2–12 Six- and 8-inch adjustable wrenches.
Courtesy Ridge Tool Co.

Many faucet handles are made of die-cast metal, whereas faucet stems are made of brass. Electrolysis, caused when dissimilar metals are in contact with each other, results in corrosion. This makes handles that become corroded difficult, if not impossible, to remove from faucet stems without dam-

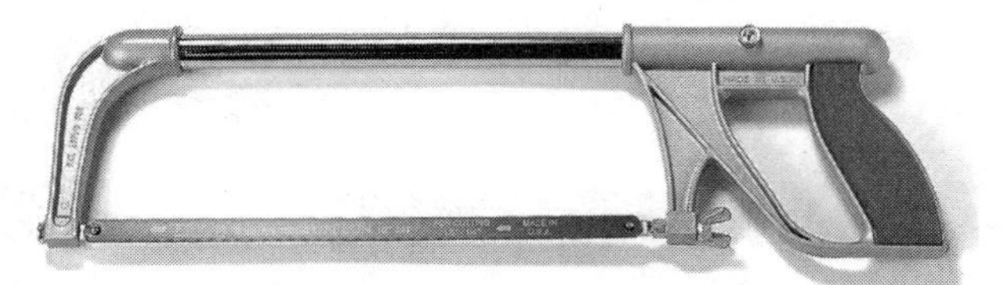

Figure 2–13 A hacksaw.
Courtesy Ridge Tool Co.

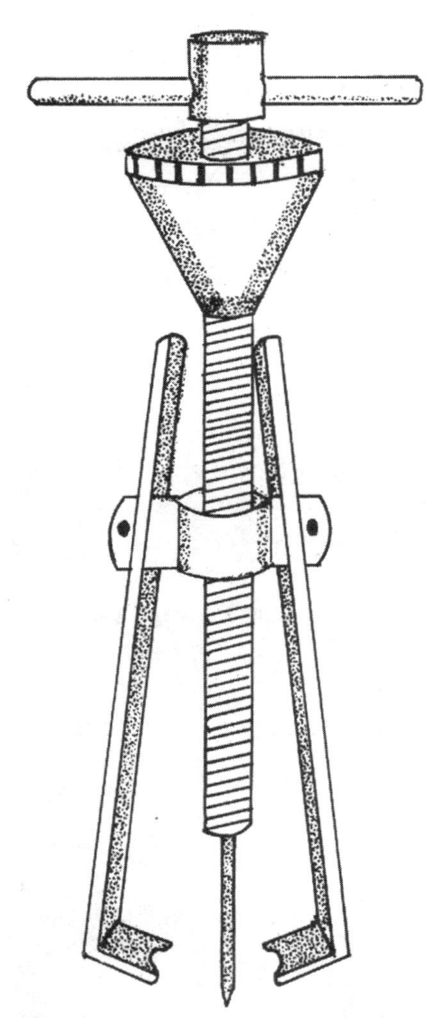

Figure 2–14 A faucet handle puller.

aging the handles. A handle puller, Figure 2–14, is the tool to use for removing faucet handles. To use it remove the index cap from the faucet handle and then remove the screw securing the handle to the stem. Insert the rod of the puller into the threaded end of the stem and place the hooks under the edge of the handle. As the tee handle of the puller is turned clockwise, the faucet handle will be pulled loose from the faucet stem.

The preceding tools will enable you to do almost all the plumbing repairs and improvements needed in the average home. If larger specialized tools such as pipe dies, pipe cutters, reamers, and pipe vises are needed, they can be rented at tool rental stores for a nominal fee.

Tools You May Need Occasionally

Closet augers, Figure 2–15, are used to remove foreign objects from a toilet bowl. The average family probably never needs one, but you may if you have small children. If you do decide to buy one, buy a good one. An inexpensive one won't do the job. Here's how to use it: Pull up on the cable until the end is against the cable housing. Then insert the cable end into the toilet trap and push down on the auger crank, turning the crank handle as you push. The cable is long enough to go completely through the toilet bowl. Objects such as a toothbrush or a pencil can become wedged in the trap and can be difficult to remove. Repeat this procedure if necessary to clear the obstruction.

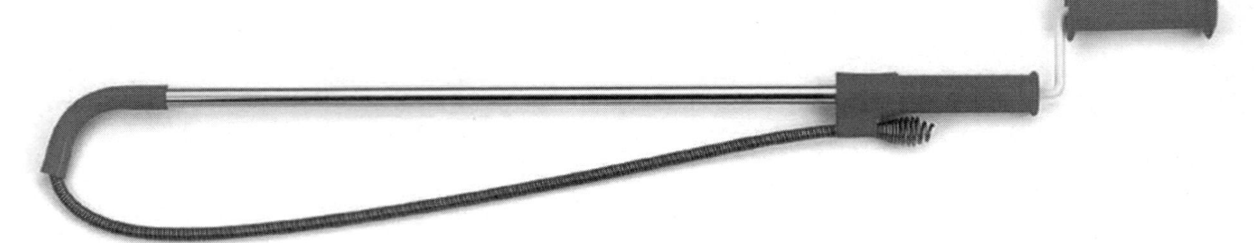

Figure 2–15 A closet auger.
Courtesy Ridge Tool Co.

CAUTION: If you think you have removed the object that caused the stoppage, remove the toilet tank lid before flushing the toilet. If, when you flush, the toilet the bowl is still stopped up, push the tank ball or flapper ball into place on the flush valve to prevent water in the tank from flowing into the bowl, causing the bowl to overflow. Then try using the closet auger again.

Pipe taps, Figure 2–16, are used to cut female threads, for cleaning rust or corrosion from female threads, and for straightening out damaged threads. Today 1/2- and 3/4-inch taps are used mostly for straightening out threads in plastic fittings.

A fine-toothed saw, Figure 2–17, makes straight cuts of plastic pipe such as PVC, CPVC, and acrylonitrile-butadiene-styrene (ABS).

A hand-spinner drain cleaner, Figure 2–18, with a 1/4-inch cable is recommended for kitchen sink and lavatory drains. Spinning the cable helps work the cable end around turns.

When PVC or CPVC tube is cut, a burr is left on the male pipe end. The tool shown in Figure 2–19 will remove this burr.

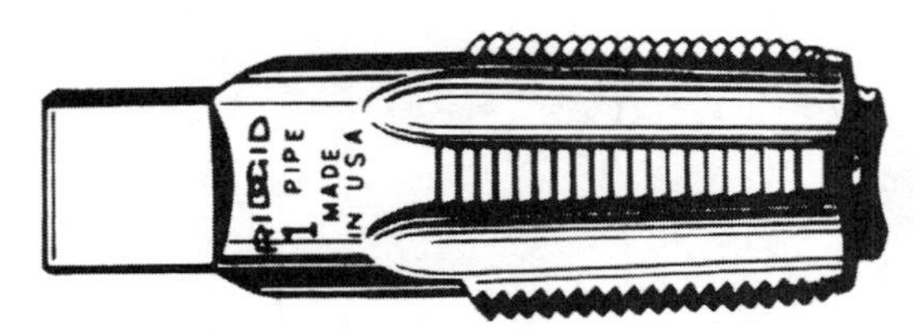

Figure 2–16 A pipe tap.
Courtesy Ridge Tool Co.

Figure 2–17 A saw will make straight cuts in PVC, CPVC, and ABS plastic pipe.
Courtesy Ridge Tool Co.

Figure 2–18 A hand-spinner drain cleaner.
Courtesy Ridge Tool Co.

Figure 2–19 A deburring tool for plastic pipe.
Courtesy Ridge Tool Co.

Figure 2–20 A torpedo level.
Courtesy Ridge Tool Co.

Figure 2–21 A Midget copper cutter.
Courtesy Ridge Tool Co.

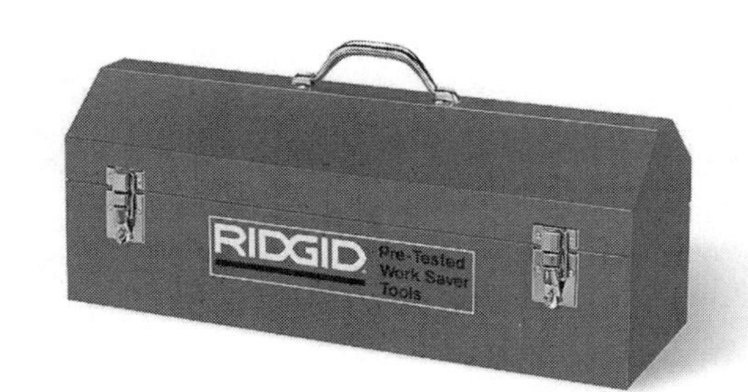

Figure 2–22 A good tool box helps keep tools organized.
Courtesy Ridge Tool Co.

A torpedo level, Figure 2–20 can be slipped into a hip pocket and be available when a pipe needs to be level.

The midget cutter shown in Figure 2–21 is especially useful when copper or plastic tubing that is near a floor, close to a wall, or in a corner needs to be cut.

Now you have a good selection of tools. These tools, which can be purchased at home improvement stores and plumbing supply houses, can last a lifetime if taken care of. The tool box shown in Figure 2–22 will help keep your tools organized and available when you need them.

3

Repairing a Control Valve

TOOLS NEEDED

6-inch adjustable wrench

8-inch adjustable wrench

MATERIALS NEEDED

00 flat faucet washer

I have made this the first repair chapter in the book because *before* you can do many of the projects in this book, *you are very likely* to have to do this one. You have bought a new kitchen or lavatory faucet or a fill valve for the toilet tank, and you're ready to install it. But first, you must turn off the water to the faucet or the toilet.

Compression stops, Figure 3–1, are the control valves, usually located below the countertop of a kitchen sink or a bathroom lavatory or below and at the left side of a toilet tank, that control the water supply to the fixture. Reach under the countertop or under the toilet tank, grasp the oval-shaped compression stop handle, and turn it clockwise (as you face it), or *try* to turn the handle.

If the handle turns until it cannot be turned any further and water does not flow or drip, either from the faucet or into the toilet tank, the valve is okay and you can skip this chapter and start installing the faucet or fill valve.

When a compression stop has not been turned off for a long time, lime or similar minerals can make the handle very hard to turn. Often, turning the stop off and on a few times will solve this problem. The handles on compression stops are made of pot metal (die cast), so be careful; if pliers are used to turn the handle, the handle may be damaged.

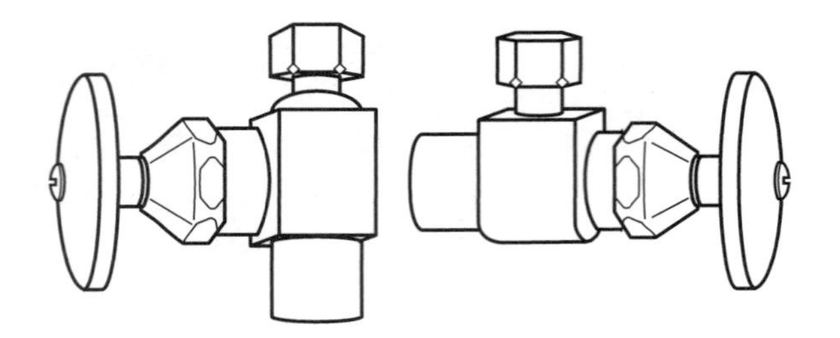

Figure 3–1 Compression stops.

If the compression stop handle, Figure 3–2, cannot be turned, the stem is either corroded or the packing under the bonnet nut has dried out, preventing the valve stem from turning. If the handle cannot be turned or if it turns until it cannot be turned any further but the water continues to flow or drip, a new 00 size *flat* rubber faucet washer must be installed on the valve stem.

The water supply to the compression stop must be turned off to repair the stop. This usually requires turning off the cold water supply to the building. The type of valve used for this purpose will depend on the regulations of the water utility. In some areas globe valves are used; in others full port valves may be used. If a wheel handle valve is used, all wheel handle valves turn clockwise to shut them off and turn counterclockwise to turn them on. Ball valves only require a quarter turn for OFF or ON action.

You may wonder why you should repair the compression stop if you have to shut off all the water in the home. There is a very good reason. If you repair or install the faucet or the toilet fill valve and a leak occurs (and it happens to the best of us), the water can be shut off *at the fixture* immediately before any damage is done. This is

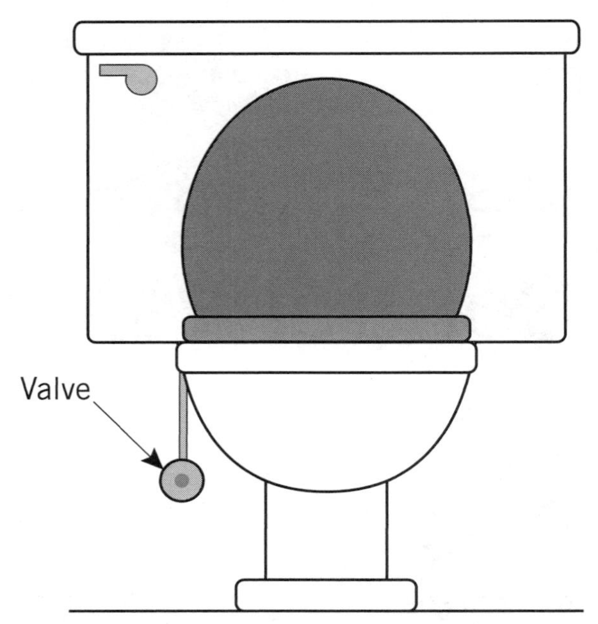

Figure 3–2 Water control valve for a toilet.

why *all* plumbing codes require a control valve or valves at every fixture.

The main valve controlling all the water to the home may be located in any one of these locations:

1. On the water main side of the meter if the meter is located in the building.
2. On the water main side of the meter in a water meter pit outside the building.
3. Where the cold-water service pipe enters the building through a basement wall or a garage wall.

When you have located the main water shutoff valve and turned this valve off, open a faucet to verify that the water is turned off.

Now we will take the compression stop apart and repair it. The handles on these valves are made of die-cast metal, are round or oval-shaped, and are brittle. If pliers are used to try to turn the handle, the handle may be damaged. To remove the stem, the bonnet nut, Figure 3–3, must be turned counterclockwise to loosen and remove it. Two adjustable wrenches should be used, one to turn the bonnet nut and the other wrench placed on the valve body to serve as a backup to prevent the valve from turning. When the bonnet nut is removed, the valve stem can be unscrewed out of the valve body.

Next a *00 flat* rubber faucet washer is forced over the stem as shown in Figure 3–3. Now the stem is repaired and the valve can be reassembled. Screw the stem back into the valve body. Place a small amount of light grease (vaseline will do) in the stem opening in the bonnet nut and start the nut on the valve body. Turn the stem handle counterclockwise two turns and then tighten the packing nut using two wrenches, one as a backup and the other to tighten the nut. When the nut is tight, turn the stem handle clockwise to turn the valve off.

Turn the main water valve back on, turn off any valves that are open, and check for leaks. If there are none, the compression stop has been repaired.

When loosening a connection to a water line, two wrenches should always be used, one wrench to loosen the desired connection and the other to prevent any movement in the part to which the connection is made.

Now you can get on with the job you wanted to do when you found that the control valve needed repair.

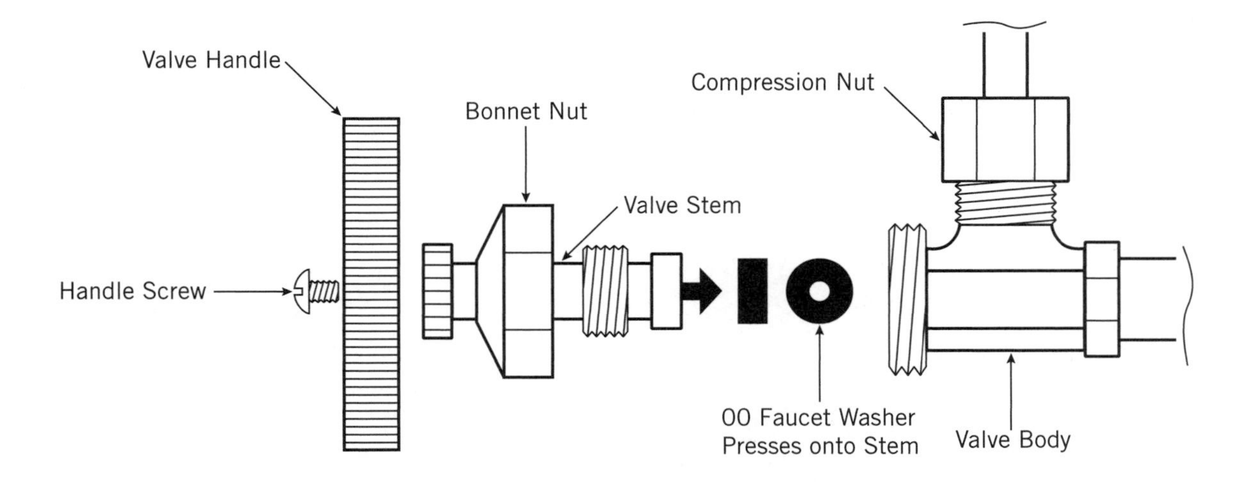

Figure 3–3 Exploded view of a compression stop.

4

Measuring and Cutting Pipe

The information in this chapter applies to steel, copper, DWV, and CPVC pipe and fittings. This chapter is presented to explain the way a plumber measures pipe. Many repair and remodeling projects require relocating or adding to existing water and drainage piping. Cutting a length of pipe only to find it is 1/2 inch too long or 3/4 inch too short is very frustrating. An inside-reading rule, Figure 2–4, or a steel tape, Figure 2–5, is indispensable for making accurate measurements. An inside-reading rule is best because it lays flat when you are marking a cut or reading a plan.

When you are measuring pipe, do it the way a plumber does. End-to-center measurements take the guesswork out of measuring. This is true whether the pipe is steel, copper, cast iron soil pipe, PVC-ABS, DWV, or CPVC. When you measure end-to-center, you always allow for the makeup of one joint. If a fitting is made (screwed, soldered, or glued on) on one end, the measurement is made from the center of that fitting to the end of the pipe, and allowance is made for makeup on the male end, the pipe will fit.

Figure 4–1 (A) shows tees added into existing piping and new piping added as connections for an automatic washer, assuming that pipes are on the same level against the floor joists on the ceiling of a basement. Because the hot and cold piping are on the same level, it will be necessary

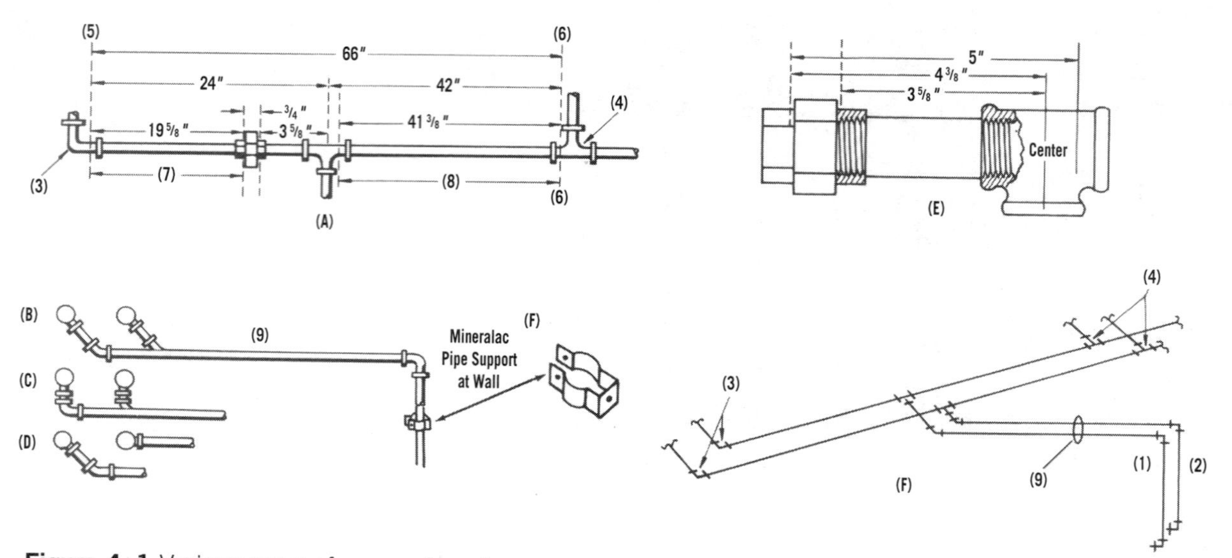

Figure 4–1 Various ways of measuring pipe.

to use fittings to enable the new pipe to pass under or over the existing pipe. Figure 4–1 (B), (C), and (D) show different ways in which this can be done.

In Figure 4–1 (B) and (C) the new piping is on the same level. In Figure 4–1 (D), the new hot-water pipe is above the cold-water pipe. In Figure 4–1 (B) and (C), the two pieces of vertical piping would be the same length. In Figure 4–1 (D) the new vertical hot-water pipe would be longer than the new cold-water pipe. If you had your own pipe dies and vise, the needed cuts and threads could be made on the job. A small job such as that illustrated would not justify the cost of renting these tools if a hardware store or plumbing shop that cuts and threads pipe is nearby.

The correct and most accurate way to measure cut lengths of pipe is to make end-to-center measurements. Figure 4–1 (A) shows a section of piping with a tee, nipple, and union installed between points 3 and 4. End to center means from the end of the pipe, with the fitting made up, or tightened onto the pipe, to the center of that fitting. To make an end-to-center pipe measurement, it is necessary to know how to measure the end-to-center distance of a fitting. Figure 4–2 (A) shows a full-size drawing of a 1/2-inch 90° elbow. X is the distance the thread makes up (screws into) the fitting, 1/2 inch. The distance Y is from the end of the thread of the pipe to the centerline of the fitting.

In Figure 4–1 (E) we see a 3-inch nipple made up into a tee. The nipple length (3 inches), plus the end-to-center measurement of the fitting (5/8 inch), gives an end-to-center pipe measurement of 3 5/8 inches. Suppose you want to cut a piece of pipe to measure 24 3/4 inches end-to-center of a tee. The actual length of the pipe would be 24 1/8 inches. The 24 1/8 inches plus the 3/8 inch (end-to-center of the fitting) would equal 24 3/4 inches from the end of the pipe to center of the tee, the desired length. Also shown in Figure 4–2 are a full-size 1/2-inch malleable tee, a 1/2-inch malleable 45° ell, and a 1/2-inch union.

Figure 4–1 (F) is an isometric drawing of the piping shown in Figure 4–1 (A).

From the front side, facing any fixture, the hot-water pipe or faucet should always be on the left side. The tees shown in Figure

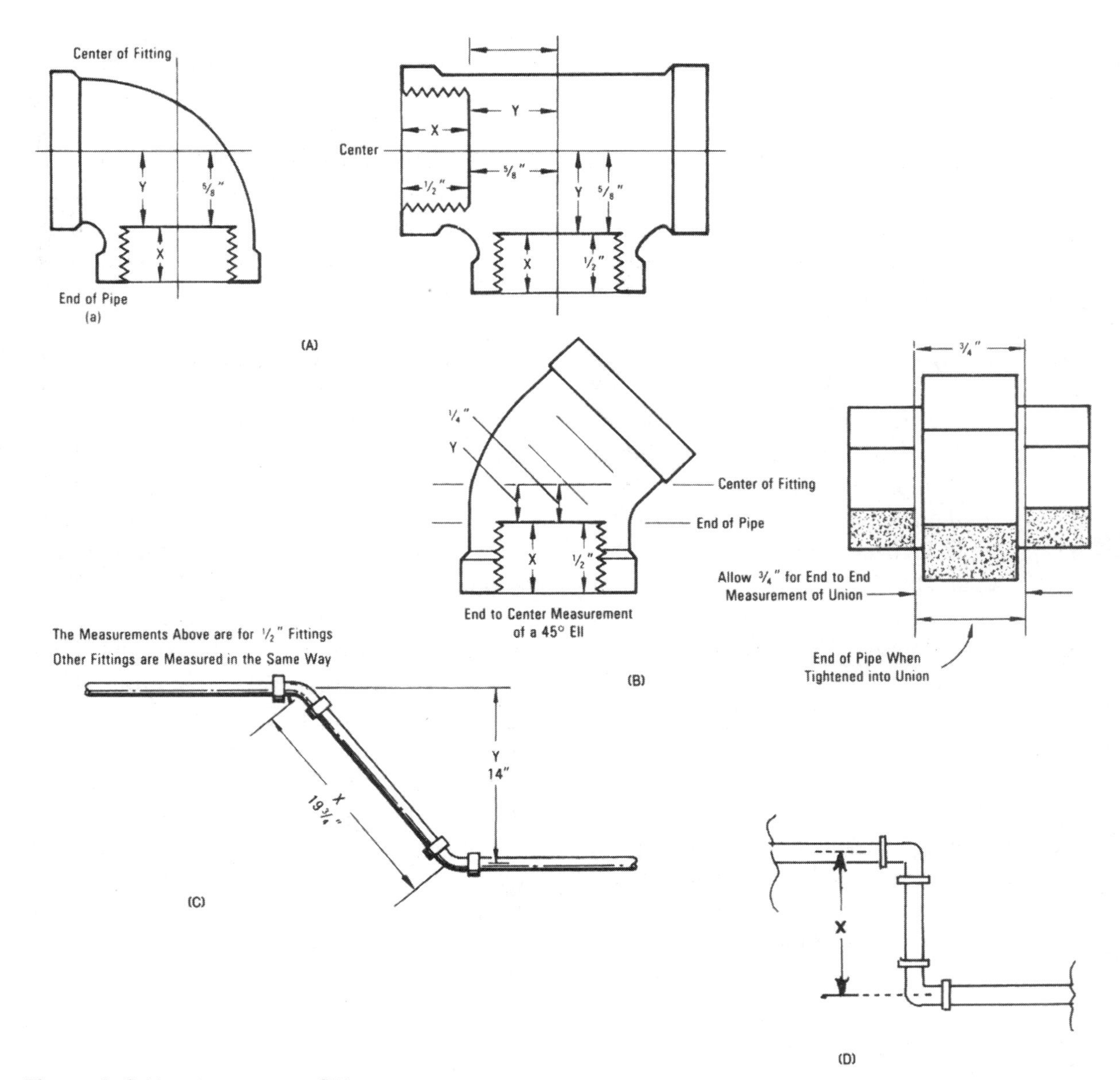

Figure 4–2 How to measure fittings.

4–1 (A) should be placed in such a manner so as to bring the piping to the desired location with the hot-water pipe on the left. When a fitting is added to piping between other fittings, 3 and 4, and 3 and 4 cannot be turned, a union must be used to join the new pipe. The distance between 5 and 6 is the end-to-end measurement of the existing piping. After a tee, a nipple, and a union are added (cut into the existing piping), this overall distance must remain the same; therefore some pipe must be cut out to make room for the tee, union, and nipple.

Figure 4–1 (A) explains how the actual cut lengths of pipe are determined. For the purpose of this explanation, the original piping length between 5 and 6 is 66 inches. If the center of the tee in the cold-water pipe is to be 24 inches from the makeup point 5, the actual cut pipe length (7) would be 19 5/8 inches. The other cut piece, Figure 4–1 (8), would then be 66 inches minus 24 inches end to center of tee: 42 inches minus 5/8 inch to makeup of tee, or 41 3/8 inches end to end.

To determine the end-to-center measurement of 9 in Figure 4–1 (F) and Figure 4–3, and using the pipe support shown, the distance from the wall line to the center of the pipe support, when the support is mounted on the wall, is 3/4 inch. If a rule is held against the wall as shown, and the end of the pipe (where the pipe makes up into the 45° elbow) to the wall measurement is 34 inches, the end-to-center measurement of this pipe is 34 inches minus 3/4 inch, or 33 1/4 inches end to center of the elbow.

The drop pieces of pipe, 1 and 2, can be cut to any desired length. All of the end-to-center measurements given in this example are for 1/2-inch pipe. End-to-center measurements for other sizes of pipe and fittings are determined in exactly the same way. The table in Figure 4–3 shows these measurements for other common pipe sizes.

It is often necessary to offset a run of pipe around a light fixture, post, or other obstruction. An offset can be made using 90° elbows, but a 45° offset is better because there is less friction (and therefore less pressure drop) when 45° fittings are used. It is very easy to calculate the length of pipe needed for a 45° offset. The formula for a 45° offset is 1.41 times the square break. The *square break* of an offset is the lateral center-to-center measurement of the offset. This is shown as X in Figure 4–2 (D).

For example, it is necessary to offset the run of 1/2-inch pipe 14 inches. The center-to-center measurement of the offset (the square break) shown as Y in Figure 4–2 (C) is 14 inches. Using the formula for figuring 45° offsets, 1.41 times the square break (14 inches) equals 19.74. Rounding off the .74 to .75 shows that the center-to-center measurement between the two 45° elbows is 19 3/4 inches. In Figure 4–2 (B) the end-to-center measurement of a 1/2-inch 45° elbow is 1/4 inch. There is a 45° elbow at each end of the offset, so we deduct 1/2 inch (1/4 times 2) from 19 3/4, and the answer, 19 1/4 inches, is the end-to-end length of the pipe in the 45° offset.

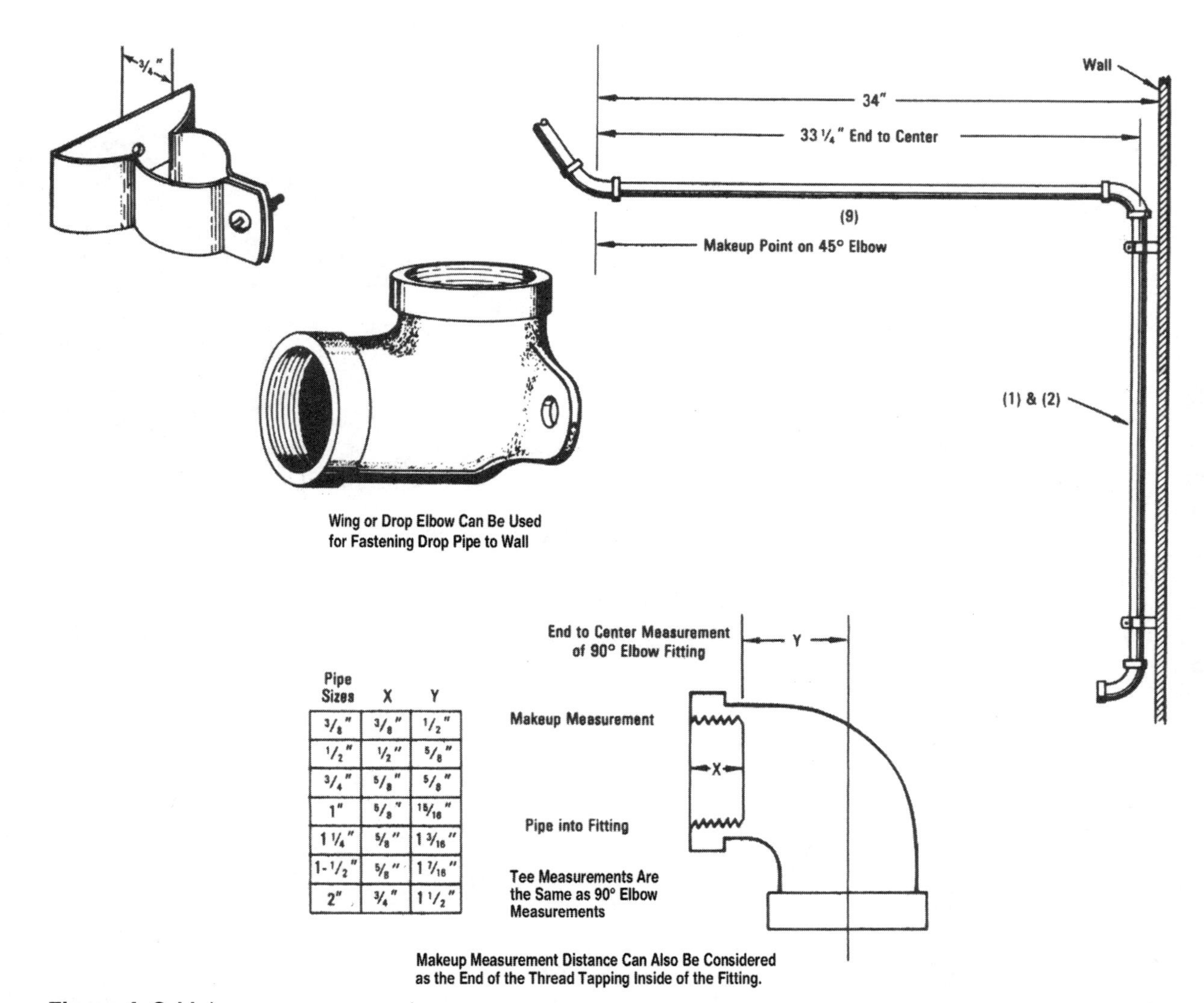

Pipe Sizes	X	Y
3/8"	3/8"	1/2"
1/2"	1/2"	5/8"
3/4"	5/8"	5/8"
1"	5/8"	15/16"
1 1/4"	5/8"	1 3/16"
1-1/2"	5/8"	1 7/16"
2"	3/4"	1 1/2"

Figure 4–3 Makeup measurements.

5

Working with Copper Tubing

Types of Copper Tubing

Three types of copper tubing are used for water piping: K, L, and M. Type K has the thickest wall and is used primarily for water service lines or in buried or inaccessible locations. It is manufactured in both soft copper rolls and in 20-foot lengths.

Type L has a thinner wall than type K and is also available in both rolls (soft copper) and 20-foot lengths (hard copper). Type L is used primarily in above-ground locations and in inaccessible locations.

Type M is made in hard copper only in 20-foot lengths. Type M is a thin-wall copper tubing and should only be used in accessible locations.

Measuring Copper Tubing

There is a right and a wrong way to measure pipe. The right way is always to make end-to-center measurements. If you are repairing or replacing a line of pipe and you want to connect to an existing line running at a 90° angle, the measurement should be made to the *center* of a 90° elbow. Or, you may want to reach a certain point and then make a 45° turn. The measurement will be accurate if you measure to the center of a 45° elbow. The end-to-center measurement of a copper fitting is very easy to figure because the fitting, whether it is a ferrule or a solder (sweat) type, is recessed for the pipe. The distance from the end of the piping recess to the center of the fitting is the end-to-center measurement of the fitting. Detailed instructions for how to measure pipe and fittings are given in Chapter 4.

Making Soldered Joints

Copper tubing and fittings can be joined in three ways:

1. Joints can be soldered.
2. Ferrule-type compression fittings can be used.
3. Flared-type fittings can be used to make joints.

There are three rules that must be followed when soldering copper tubing:

1. The male ends of the pipe and the female ends (sockets) of the fittings must be clean and bright.
2. The joints must be dry.
3. Heat must be applied at the right places. Capillary action will then pull the solder into the joint. If you learn these rules and follow them and apply heat to pipe and fittings as shown in Figure 5–1, you will make perfect solder joints every time.

It has been proven that water passing through soldered joints made with lead-type solders will leach out (dissolve) the lead in the solder. This lead is then present in the water where it can be consumed in cooking and drinking.

The Safe Drinking Water Act, *Public Law 99-339,* prohibits the use of solders and fluxes having a lead content in excess of 0.2%, (two tenths of 1%). When making joints or fittings in any public or private potable water system, 95/5 solder (95% tin and 5% antimony) meets the requirements of lead-free solder and can be used to make joints in potable water systems.

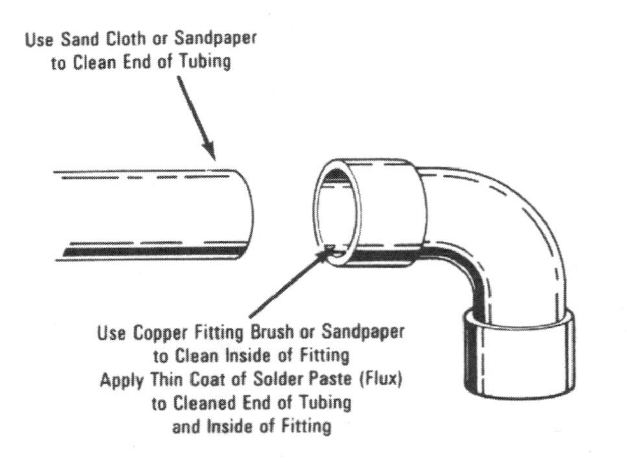

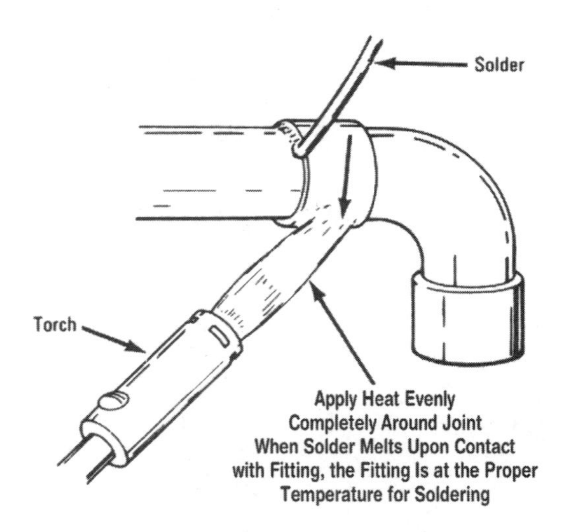

Figure 5–1 How to make a good solder joint.

A section of piping to be repaired is shown in Figure 5–2. When heat is applied to solder a copper joint, air in the pipe is heated, creating pressure in the pipe. When a repair must be made, the first step is to turn off the valve controlling that section of pipe. Then, *a valve or faucet downstream of the repair* must be opened to relieve any pressure that will build up in the pipe. The built-up pressure, if not relieved, will cause the solder to be blown out of the joint, leaving a hairline crack or pinhole that will leak when water is turned on.

Clean the male ends of the pipe and the socket ends of fittings with sand cloth or fine sandpaper. When clean, they will be bright and shiny. Then coat the cleaned ends with solder flux, Nokorode or Oatey's, using an acid brush. Join the cleaned pieces and apply heat with your torch as shown in Figure 5–1. Heat should be applied at the base of the fitting socket because heat applied here will cause capillary action to pull the solder back into the joint. Plain 95/5 solder must be used, *not* acid- or rosin-core solder. Unroll 8 or 10 inches of solder and apply the end of the solder to the joint. As soon as the solder starts to melt, apply the flame to the center of the fitting and then all the way around the fitting. Follow the flame with the solder end. When solder runs all the way around the joint, the repair is made.

A solder joint can be made in any position, right side up, upside down, horizontal, or vertical. If heat is applied at the right places, capillary action will draw the solder around the joint.

When soldering copper tubing joints, the pipe and fittings must be dry. It is impossible to solder a joint that has water in it. Water will turn to steam and prevent the solder from melting. If you are trying to solder a joint and find that water will not shut off completely, allowing a trickle of water to flow, use this old plumber's trick. You can stop this trickle long enough to solder a

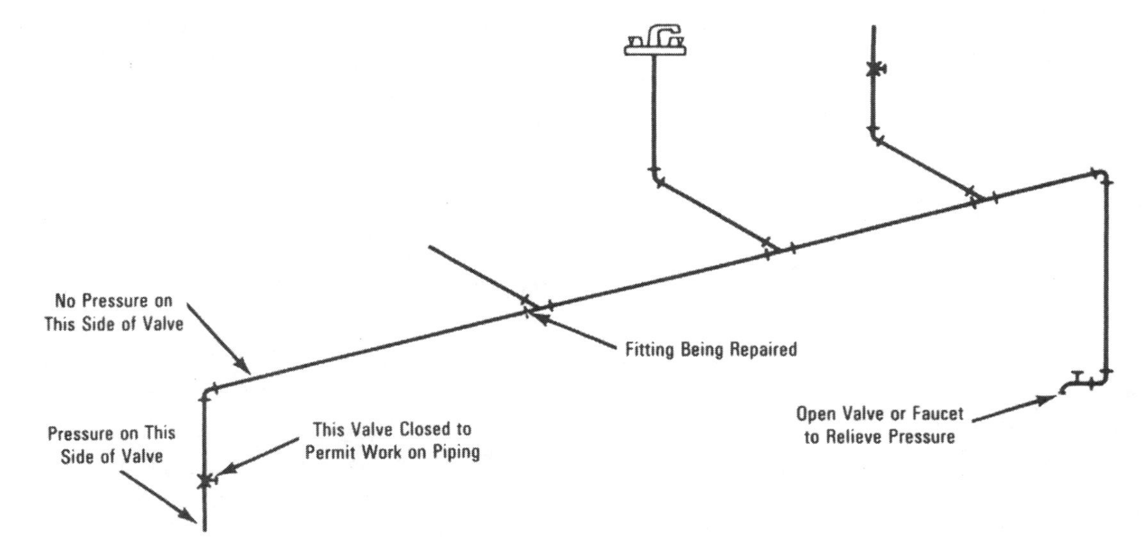

Figure 5–2 Repairing a leak in a section of pipe.

joint if you stuff plain white bread (remove the crust) into the pipe and pack it tight with the eraser end of a pencil as shown in Figure 5–3. When the solder joint has been made and water pressure has been turned on, the bread plug will disintegrate and can be flushed out of the pipe through an open valve or faucet. If the faucet has an aerator on the spout, remove the aerator while flushing out the piping.

Solder Fittings

Pictured in Figure 5–4 are some common solder fittings. The 90° ell (elbow) has two female (socket) ends, and the 90° *street* ell has one male end and one female socket. A street fitting is any fitting that has one male end and one or more female ends. Street fittings made of cast iron or malleable iron *restrict* flow; copper street fittings offer no restriction to flow. Tees and ells are available both as straight pipe-size fittings and as reducing tees and ells. A tee is always read (described) end 1, end 2, side 3. In the following example × means by. A tee with all openings the same size (1/2 × 1/2 × 1/2) is a 1/2-inch tee. A tee with the same size openings on both ends and a different size opening on the side (1/2 × 1/2 × 3/8) is called a 1/2 by 3/8 tee. Knowing how to ask for a fitting is the way to be sure you get the right fitting.

Copper-to-iron adapters are used when connecting copper tubing to pipe threads. The term *iron* does not necessarily mean connecting to steel or wrought iron pipe; the connection may be to a threaded valve or fitting. It means that the copper fitting is made with either a male or female pipe thread. A copper-to-male iron pipe (M.I.P.) adapter is shown in Figure 5–4. The fitting is made of copper with a female socket on one end and a male pipe thread on the other. A copper-to-female iron pipe (F.I.P.) adapter is also shown in Figure 5–4. When a solder joint is to be made using an adapter fitting, care must be taken not to let solder run into the pipe threads.

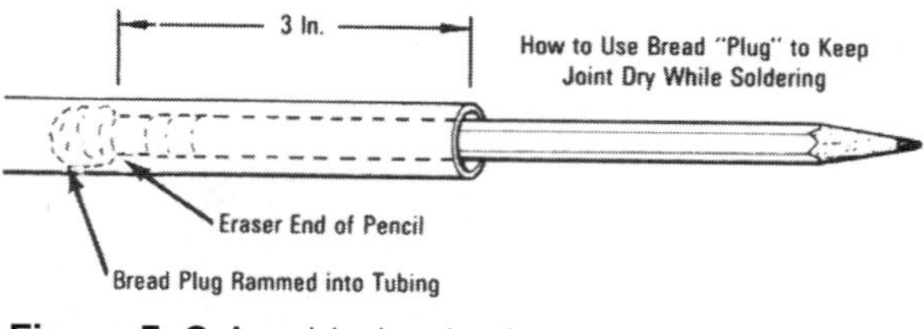

Figure 5–3 An old plumber's secret.

Compression Fittings

Compression fittings can be used to make repairs or insert fittings or valves into existing piping. Solder is not used with compression fittings. There are two types of compression fittings: ferrule and flare. When a ferrule-type compression fitting is used, the compression nut is inserted onto the pipe, followed by the ferrule, as shown in Figure 5–5.

The pipe is then inserted into the fitting, in this case a 90° elbow. As the compression nut is tightened on the fitting threads, the ferrule is compressed against the pipe, sealing the joint. When a 3/8-inch chrome-plated soft copper flexible tubing is used for water supply connections to faucets and fixtures, the tubing must usually be bent or offset. This may kink the tubing or make it out-of-round, making it impossible to slide the

90° Elbow

90° Street Elbow

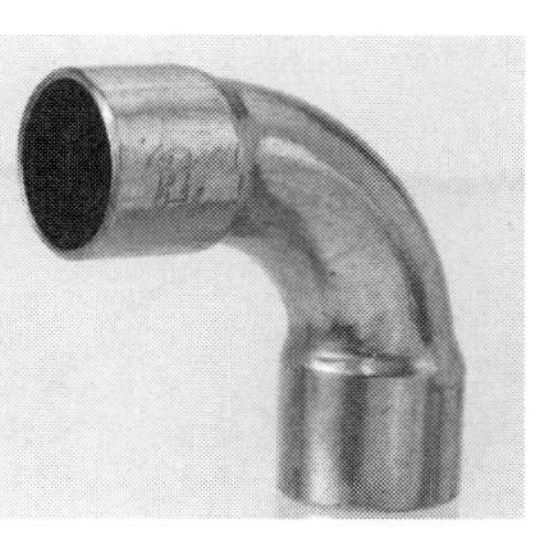

Long-Turn 90° Elbow

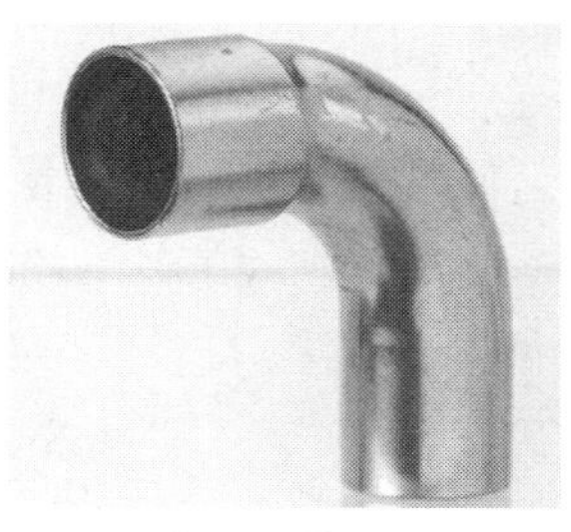

Long-Turn
90° Street Elbow

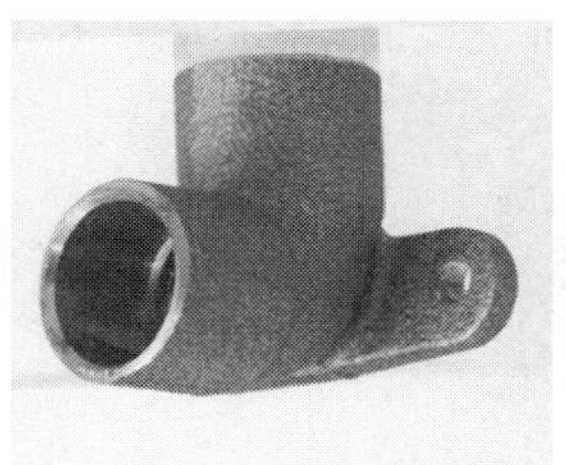

Wing (Drop) Elbow

c × F.I.P.
Wing (Drop) Elbow

Wing (Drop) Tee

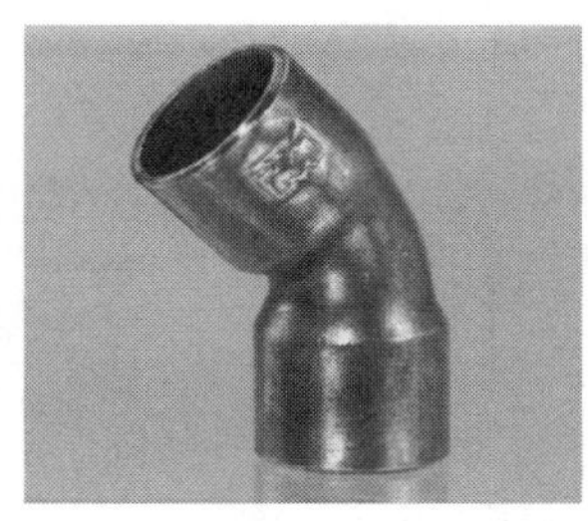

45° Elbow

Union

Male c × F.I.P. Adaptor

Bushing

c × F.I.P. × c Tee

Coupling

Fitting Reducer

c × F.I.P. Adaptor

Figure 5–4 Common copper fittings.
Courtesy Mueller Industries

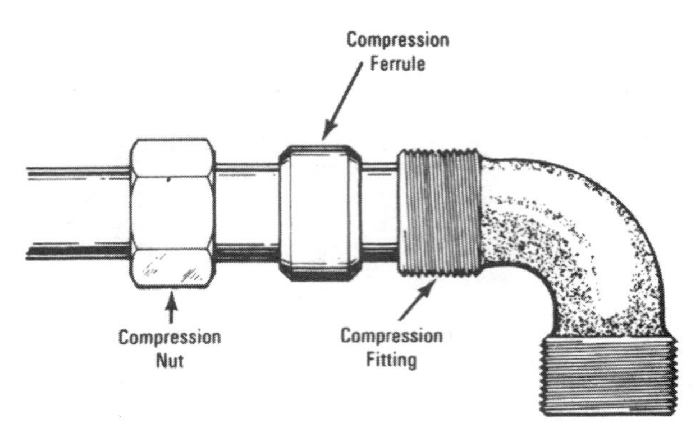

Figure 5–5 A ferrule-type compression fitting.

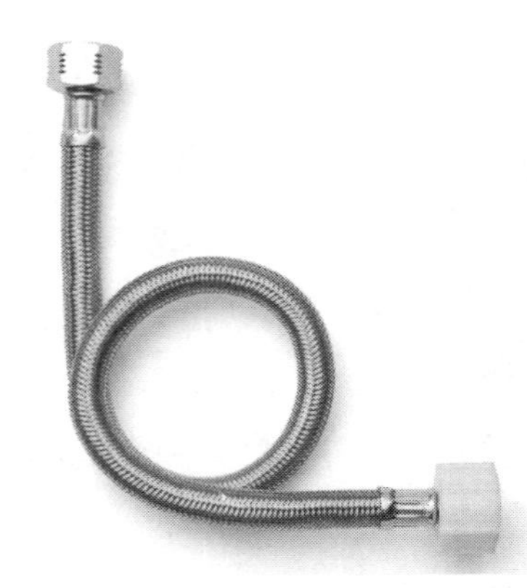

Figure 5–6 A braided flexible fixture connection.
Courtesy Fluidmaster

compression nut and ferrule onto the pipe. Therefore, a new product, a braided stainless steel flexible supply connection, Figure 5–6, is best for making these connections.

Flare-type compression fittings are often needed when installing soft copper water service piping. Hard copper tubing cannot easily be flared; if hard copper tubing must be flared, it must first be annealed or softened. Hard copper can be annealed by heating it to a cherry red, and then allowing it to cool.

Copper flaring tools are available at plumbing supply stores. The tool used for flaring 3/4- or 1-inch soft copper tubing for water services is a swedging tool. The tool is inserted into the tubing and the flare is produced when the tool is driven into the tubing with a hammer.

A flaring tool used for smaller soft copper tubing is shown in Figure 5–7. This tool is called a flaring block. It will produce a perfect flare.

Figure 5–7 A copper tubing flaring block.
Courtesy Ridge Tool Co.

How to Stop Water Hammer

An air chamber (cushion) fitting, Figure 5–8, is used to prevent water hammer, the hammering noise often heard when a faucet

Figure 5–8 An air-cushion fitting.
Courtesy Watts Regulator

is closed very quickly. When water being forced by pressure through piping is suddenly stopped, the flow is instantly reversed, causing the hammering noise. Solenoid valves on clothes washers and dishwashers close very quickly, often causing water hammer. An air cushion fitting will hold a pocket of air in the top of the fitting, providing space for the expansion of the suddenly reversed flow of water. An air cushion fitting installed in a tee or at the end of a run of piping will prevent water hammer. An air cushion fitting must always be installed vertically.

6

The Thieves Hiding in Your Toilet Tank

TOOLS NEEDED

Large slip-joint pliers

Adjustable wrenches

Large sponge

Everything considered, the maze of wires, floats, rods, levers, tubes, guides, and balls shown in Figure 6–1 (allegedly invented by the legendary Rube Goldberg) work amazingly well. Eventually mineral buildup, Murphy's Law, and normal wear and tear take their toll and cause these parts to malfunction. When this happens, they will cost you money. Before we go any further, if you've ever been told to put bricks and/or bottles in your toilet tank to save water, *remove them now.* Nothing belongs in a toilet tank except the working parts made for the tank.

Products designed to be placed in toilet tanks to clean the toilet bowl with every flush are sold in grocery and hardware stores. If these products contain chemicals strong enough to clean toilet bowls, they do not belong in toilet tanks. They can be very dangerous. (See Chapter 9, "Protecting Your Family's Health.")

There are two ways to do the work described in this chapter. The first and best way is explained in Part 1 wherein we put new parts in the tank.

But my editor is a very particular person and he said, "Charlie, if this tank has been working all this time, rather than replace almost everything, why not just fix the parts that are bad?" So I've divided this chapter into two parts. Part 1 is the way I would do it for myself. If it is done as explained in Part 2, where we see a problem and fix just that one problem, it's usually a temporary fix but it *is* a fix. So, you decide what you want to do. You have two choices:

1. Take a look at Figure 6–1. You can throw most of those parts away, put in new parts as explained in Part 1, and end your problems.
2. You can replace one or more of the parts shown and hope the problem's solved.

Let's hope you choose to follow Part 1. If you do, you will scrap most of the parts and replace them with new parts that will give you years of trouble-free service.

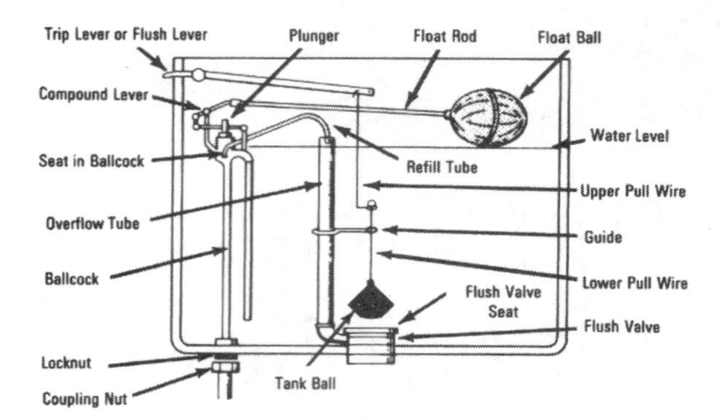

Figure 6–1 Old fashioned tank components.

Figure 6–2 A Fluidmaster 400A fill valve.
Courtesy Fluidmaster

Part 1

Here's what you will need:

- A *fill valve,* Figure 6–2. Notice that the new valve is called a fill valve, not a ballcock. The new fill valve has no float ball and float rod.
- A *braided stainless steel flexible supply,* Figure 6–3. Before you buy this little jewel, measure the distance between the top of the shutoff valve and the bottom of the toilet tank. Allow a little extra if the existing supply tube now in use is offset. Braided stainless steel flexible supply tubes are made in several lengths. If the flexible supply tube is a little long, it can be easily shaped (offset) to fit. If you buy this flexible stainless steel supply once, you'll never go back to the old chrome-plated copper or plastic supplies. Just specify what you want to use it on: faucets, toilets, water heaters, or whatever. They are made in various lengths and come with coupling nuts to fit particular uses.
- A *trip (flush) lever,* Figure 6–4.
- A *flapper-type tank ball,* Figure 6–5.

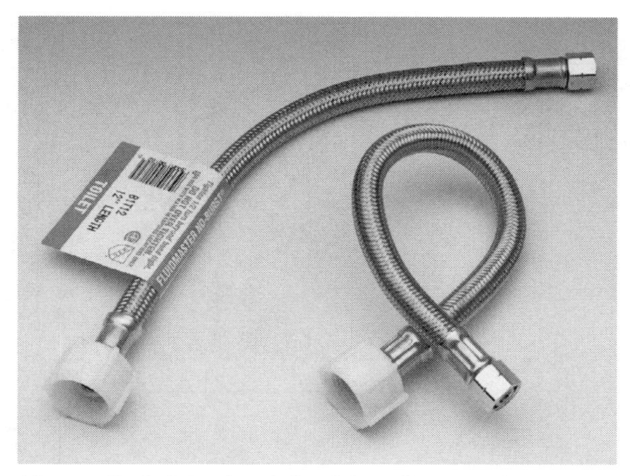

Figure 6–3 A braided flexible toilet supply.
Courtesy Fluidmaster

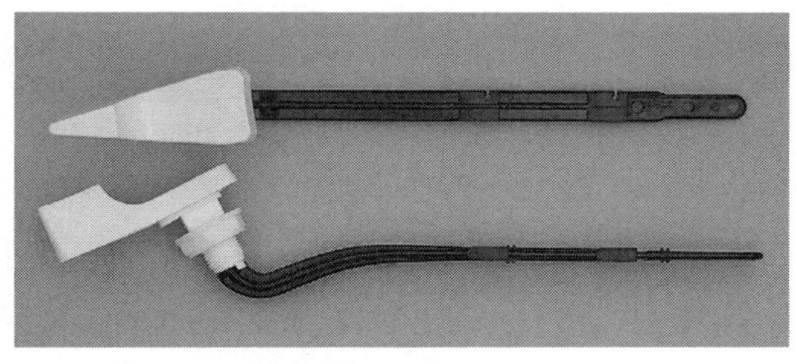

Figure 6–4 A toilet tank trip lever.
Courtesy Fluidmaster

Figure 6–5 A flapper-type tank ball.
Courtesy Fluidmaster

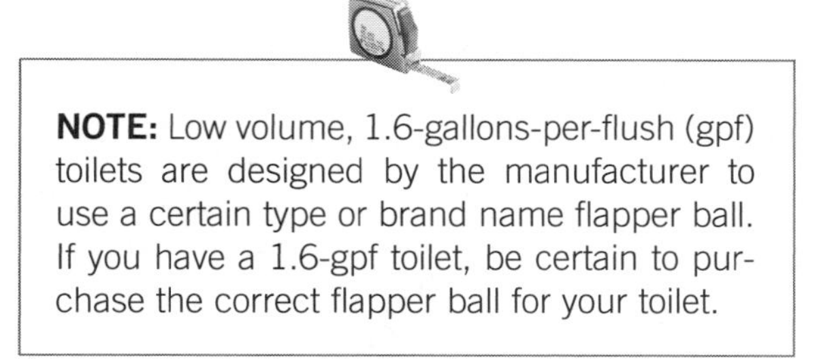

NOTE: Low volume, 1.6-gallons-per-flush (gpf) toilets are designed by the manufacturer to use a certain type or brand name flapper ball. If you have a 1.6-gpf toilet, be certain to purchase the correct flapper ball for your toilet.

These parts are available at your local home improvement store.

The fill valve you select should be an antisiphon type. (See Chapter 9, "Protecting Your Family's Health.") Plumbing codes specify that only antisiphon types be used. They are designed to prevent backsiphonage of water from the toilet tank. This information will be clearly marked on the box in which the fill valve is packed.

I recommend Fluidmaster 400A for these reasons:

1. Fluidmaster signals audibly if there is a water leak in the toilet tank. You will hear it refill the tank and shut off *when the toilet has not been used.*

2. Fluidmaster 400A is plumbing code-compliant. It is designed to prevent back-siphonage of toilet tank water into the building drinking water system. *This is very important.*

3. Fluidmaster provides years of trouble-free service.

The first step is to shut off the water supply to the toilet. The valve controlling water to the toilet normally is under the left side of the toilet tank, 8 or 10 inches above the floor, Figure 6–6. Turn this valve handle clockwise to shut off the water. If the valve handle will not turn or if water continues to run into the tank after you flush the toilet, try turning the shutoff valve off again. If water continues to run or drip into the tank, the shutoff (control) valve should be repaired. This is explained in Chapter 3.

Flush the toilet and use a large sponge to mop up the water remaining in the tank. Unscrew the old tank ball from the lower pull wire, turning the ball clockwise. Remove the upper pull wire from the flush lever, lift out the two pull wires, and discard them.

The old flush (trip) lever, Figure 6–1, is secured to the tank by a locknut on the inside of the tank. This locknut has a *left-hand* thread and must be turned clockwise, looking at it from inside the tank, to loosen and remove it. Take out and discard the old flush lever. If a pull wire guide is mounted on the overflow tube, loosen the screw securing it to the overflow tube and discard the guide.

The existing supply tube will be connected to the old ballcock by a large coupling nut, Figure 6–1. Using the large slip-joint pliers, turn this coupling nut counterclockwise (looking at it from below) to loosen and remove it. A locknut on the ballcock shank directly above the couplIng nut secures the ballcock to the tank. Turn this locknut counterclockwise (looking at it from below) to loosen and remove it. Then, lift the ballcock out of the tank.

A new flapper-type tank ball, Figure 6–5, should be installed now. The overflow tube is part of the flush valve, and the flush valve may be made with hooks projecting at the sides of the valve. The new tank ball is made to be used either by mounting the ball on the hooks or by sliding the mounting ring over the overflow tube. Directions are on the card containing the new ball.

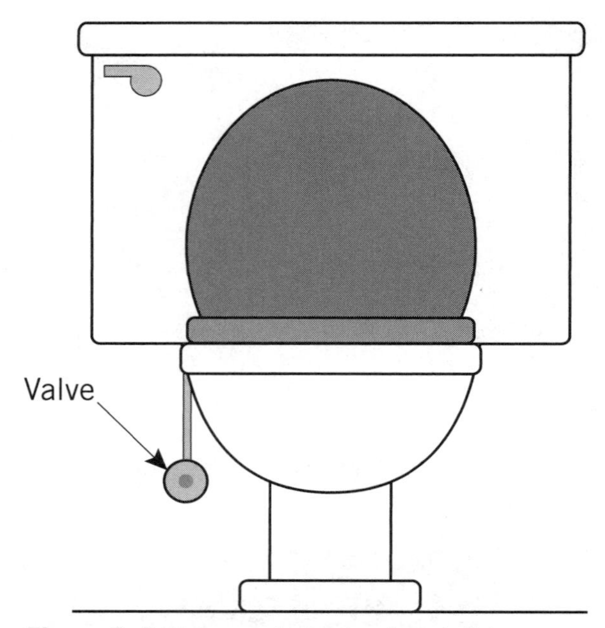

Figure 6–6 Water control valve for a toilet.

CAUTION: It has been reported that 1.6-gallon toilet tanks can malfunction if the wrong tank ball is used to replace the original. To avoid this problem *use only the original brand name tank ball.*

Remove the locknut from the new trip lever and insert the trip lever into the tank, position the trip lever so that the end is about 2 to 3 inches below the top of the tank, and start the locknut on the lever. Remember, the threads on the trip lever are left-hand threads; turn the locknut clockwise, looking at it from the inside of the tank, and tighten it one-quarter turn past hand tight. Do not overtighten the locknut; you could crack the tank.

Connect the chain on the new tank ball to the end of the flush lever. Leave some slack in the chain to allow the arm of the flush lever to raise approximately 1 inch before the tank ball is lifted.

Before mounting the new fill valve in the tank, read the instructions for adjusting the fill valve length. It is adjustable from 9 to 14 inches to fit any size tank. Read and follow the instructions for adjusting the fill valve to the correct length to fit your tank. Figure 6–7 shows how to grasp the valve. Twist the shank in or out of the valve body to adjust the valve to the correct length. At the top of the fill valve you will see the letters CL (critical length). When the fill valve is set at the correct length, the letters CL will be at least 1 inch above the top of the overflow tube.

Carefully separate the cone washer from the center of the shank washer. Slide the shank washer onto the threaded fill valve shank with the flat surface against the valve. Set the fill valve into the mounting hole in the tank with the actuating wire pointing at the overflow tube, and start the locknut on the fill valve shank. With the fill valve in the correct position, hold down on the shank and tighten the nut one-half turn beyond hand tight.

Insert the bare end of the refill tube over the serrated opening on the top of the fill valve. If there is a cap on top of the overflow tube, remove it and insert the metal clip on the other end of the refill tube onto the top of the overflow tube, *leaving a gap between the end of the refill tube and the top of the overflow tube.* The gap is to prevent the possibility of back-siphonage. *If the refill tube is inserted into the overflow tube below the water line, water will siphon out of the tank.* The refill tube may need to be trimmed to avoid kinking it.

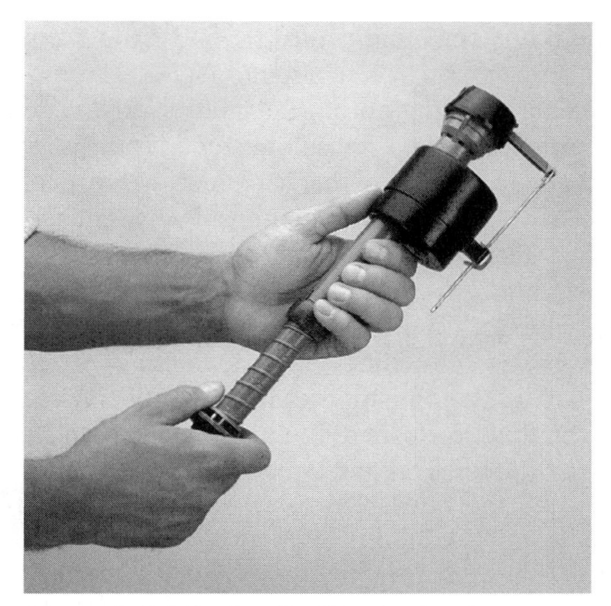

Figure 6–7 Adjusting the fill valve length.
Courtesy Fluidmaster

The new braided flexible stainless steel supply has a large coupling nut on one end and a small coupling nut on the other end. Back in the dark ages when I started to learn the plumbing trade, connecting the water supply to a toilet tank was a job. No more. The flexible tubing is long enough that it can be shaped, making an offset if needed, as shown in Figure 6–8, to connect the control valve to the fill valve. Connect the flexible supply to the fill valve first, using the large coupling nut. Turn the nut clockwise (viewed from below) and tighten the nut one-quarter turn beyond hand tight, using your large slip-joint pliers. Start the small coupling nut on the control valve, and turn the nut clockwise three or four turns by hand to make sure that it's not crossthreaded. Then tighten it with an adjustable wrench.

Turn the handle on the control valve counterclockwise to open the valve. Water will start running into the tank.

The refill tube plays a very important part in the operation of the fill valve. When the toilet is flushed, the tank ball is lifted off the flush valve, and water and waste matter leave the bowl. The bowl is emptied, and the tank ball drops down onto the flush valve preventing any more water from leaving the bowl. Here is where the refill tube comes into play.

The toilet bowl is made with a built-in trap, and the trap is filled with water each time the toilet is flushed. The purpose of the trap is not to catch anything; it is designed to prevent sewer gas from entering the building through the toilet bowl.

As the toilet tank refills through the fill valve, water is also entering the bowl through the refill tube into the overflow tube in the tank. Water continues to run into the tank through the overflow until it reaches the preset level in the tank. At that point water flow into the tank stops.

Adjust the water level by sliding the clip up or down on the fill valve. When the tank is full, the correct water level should be about 1/2 inch below the top of the overflow tube.

Congratulations. You've rebuilt the works in a toilet tank, installed a new fill valve, a new flexible supply tube, new tank ball, and a new flush lever. And, you've saved the cost of this book in the process. Not a bad morning's work.

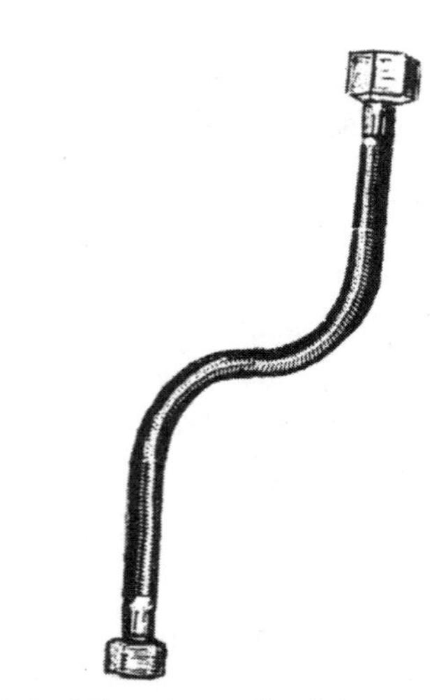

Figure 6–8 Offsetting a flexible supply.

Part 2

Let's study the chain of events set in motion when the trip lever in the tank is pressed down. Refer to Figure 6–1 as we go along. Pushing down on the trip lever pulls up the upper pull wire, which is linked to the lower pull wire. The lower pull wire, sliding through the guide, is screwed into the tank ball. The tank ball, being pulled up, becomes free of the water pressure that had been holding it in place on the flush valve seat and rises. Water then starts to rush out of the tank and into the bowl through the flush valve. As the water level drops, the weight of the float ball and the float rod exerts pressure to raise the plunger from the ballcock washer seat. The ballcock, a water valve, is thus turned on and water begins to flow into the toilet tank. The water coming into the tank flows in more slowly than the water flowing out; therefore, the tank ball drops with the receding water and is guided into proper contact with the flush valve seat by the guide. When it is correctly seated, the tank ball prevents any more water from leaving the tank.

Now that we know the sequence of events that flush out the toilet bowl, let's see some of the things that commonly go wrong in the operation of the toilet tank.

COMMON PROBLEMS AND CURES

PROBLEM

Pushing down on the flush lever does not lift the tank ball.

STOP

CAUTION: The trip (flush) lever is secured to the tank by a locknut. The locknut has a left-hand thread. Loosen the nut by turning it counterclockwise looking at it from inside the tank. Tighten the nut by turning it clockwise looking at it from inside the tank.

CURE

1. Loosen the locknut on the flush lever, inside the tank. Reposition the lever, raising the end connected to the upper pull wire slightly. Then tighten the locknut. Do not overtighten the nut; the tank is porcelain and can break if the locknut is overtightened.
2. Replace the trip lever. Wear has caused too much play in the old lever.

The ballcock will not shut off. Water rises too high in the tank and runs into the overflow tube.

1. The float ball may be rubbing on the side of the tank so that the ball cannot rise with the incoming water. Bend the float rod sideways until the float rides free on the water.
2. The float rod may allow the float ball to ride too high, causing the water to continue to flow into the tank. Using both hands, Figure 6–9, grasp the rod firmly and bend the float end down. This will cause the ballcock to shut off at a lower water level.

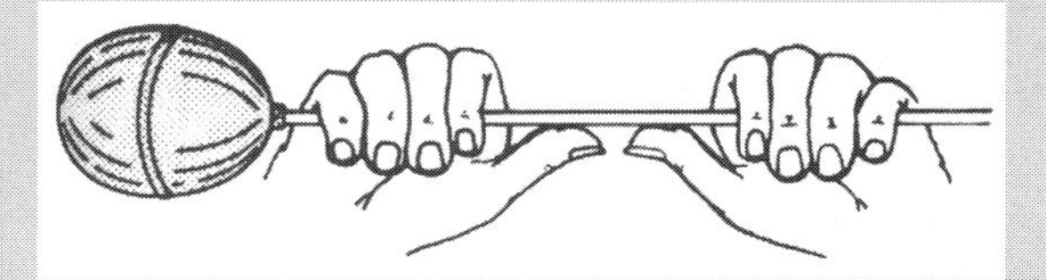

Figure 6–9 Bending a float rod.

3. The float ball may be waterlogged. Get a new float ball, lift up the float arm, turn the float ball counterclockwise to loosen and remove it, and screw the new ball on the rod by turning the float ball clockwise.

You have a high water bill and suspect a leak at the toilet tank, but you can't find it.

Pour a teaspoonful of blue ink or food coloring into the tank, wait 3 or 4 minutes, and then observe the water in the bowl. If the water has changed color, the leak is where the tank ball seats on the flush valve. Tank balls rot, deteriorate, and fail to seat on the flush valve correctly. If the bowl water changes color, a new tank ball should be installed. In most cases a flapper-type tank ball can replace a pull wire-type ball. Instructions come with the new ball.

(continues)

COMMON PROBLEMS AND CURES (CONTINUED)

CAUTION: If your toilet is a low volume flush type, (1.6 gallons per flush), use only the *exact brand name type of flapper valve that came with the toilet* as a replacement.

PROBLEM

The tank ball is okay but does not seat correctly. Water continues to run into the tank through the flush valve.

CURE

The upper or lower pull wires may be bent or the guide may not be positioned correctly to direct the tank ball onto the flush valve seat. Straighten the pull wires and reposition the guide.

7

Replacing Sink and Lavatory Traps

TOOLS NEEDED

Fine-toothed hacksaw blade

Small (1/4 inch) cold chisel

Hammer

MATERIALS NEEDED

A new trap

A trap adapter

Teflon® tape

Not too many years ago replacing a kitchen sink trap or a lavatory trap was a difficult job. Kitchen sink traps were made of cast iron, the fittings were also cast iron, and large pipe wrenches were needed (as well as a hammer to break the cast-iron fittings if all else failed). Now, thanks to copper (DWV) and PVC-DWV pipe and fittings, those days are gone, at least in fairly new homes.

Kitchen Sink Traps

If you happen to have an old home with a old cast-iron trap (called a B & F trap), you may still have to resort to the wrench and hammer method to remove the old piping. But once it is out (plumbing code in your area permitting), you can change over to PVC-DWV pipe and fittings. Then if in the future a leaking trap must be replaced, the job can be as simple as loosening and removing two slip nuts, inserting a new trap, and tightening the slip nuts.

Figure 7–1 (A) shows a trap connected to the drainage piping through a PVC-DWV compression-to-female slip adapter. Several possible piping arrangements are shown in Figure 7–1. Figure 7–1 (B) shows a compression-to-F.I.P. adapter that could be used. Figure 7–1 (C) shows a compression-to-male slip adapter, and Figure 7–1 (D) shows

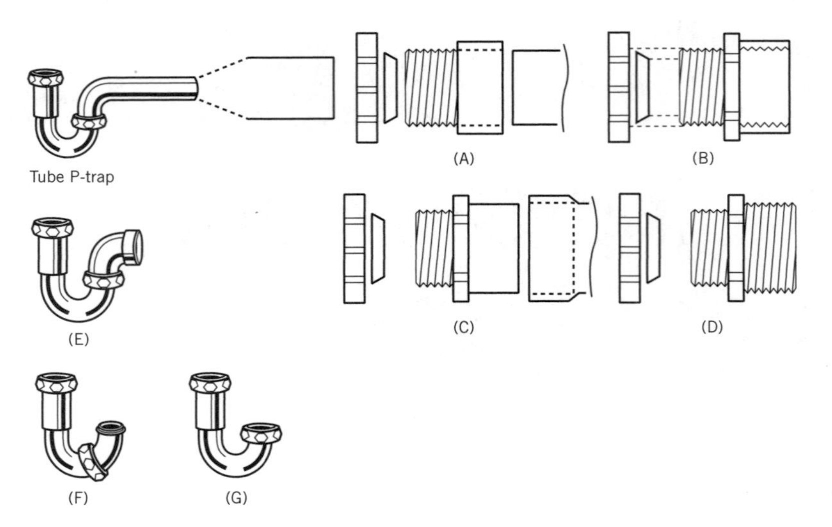

Figure 7–1 Connecting a trap with a trap adapter.

a compression-to-M.I.P. adapter that could be used. Adapters to fit galvanized pipe, copper tubing, or PVC drainage piping are also available at most home improvement stores. Figure 7–1 (E) shows a P-trap with a cast brass elbow. This type trap is made with threads on the inside of the elbow to fit a male pipe thread. To replace this type trap, the adapter in Figure 7–1 (B) and the tube P-trap would be used.

Quite often a cracked or leaking J-bend of the type shown in Figure 7–1 (F) and (G) can be replaced without changing any of the drain piping. Replacement J-bends are readily available wherever plumbing supplies are sold.

Lavatory Traps

Replacement of a lavatory P-trap can sometimes be a real problem because of the several ways traps can be connected to the drainage system. The method that was used in a particular location will depend on the age of the home and the plumbing code in effect at the time the home was built.

Until fairly recently in some areas, lead pipe was used for drainage piping and brass traps were soldered into the lead pipe at the wall line. Replacing a trap that was installed in this way is a job for a *highly skilled* craftsman. If a trap that needs to be replaced is soldered to a lead drainage pipe, now is the time to replace not only the trap but also the lead pipe. Preferably, if codes permit, replace the lead with PVC.

Galvanized steel drainage pipe is found today in many older homes. Brass tube traps were often used in these homes. A fitting called a *solder bushing* was used to make the connection. Solder bushings were made of brass, with either an 1 1/4- or 1 1/2-inch male thread (depending on the drain piping size) on the outside and with a smooth inside surface. The chrome-plated brass tube trap had to have about 1/2 inch of the chrome filed off the end of the trap arm, and the trap arm could then be soldered into the brass bushing. The trap was then screwed into the drainage fitting at the wall line. But what happens when the trap has to be replaced? Let's do it. You are making some changes in your bathroom, and you have to replace the lavatory trap. Well, now the fun begins. Slide the chrome-plated escutcheon back on the trap arm, away from the wall line. The trap arm is soldered to a solder bushing, Figure 7–2.

You try to turn the trap and the thin brass crumples. Using a hammer and small cold chisel, cut through the trap arm around the solder joint as shown in Figure 7–3.

Now the solder bushing must be removed. But, there is no way to get hold of it to unscrew it, so we'll have to cut a notch out of it and then knock the rest of the bushing out. Your hardware or home improvement store tool section will have a tool that will hold a hacksaw blade, leaving several inches of the blade open, or you can wrap some duct tape around the teeth on a fine-toothed hacksaw blade to protect your hand and make two cuts in the solder bushing as shown in Figure 7–4.

Try not to cut too deep into the threads in the fitting. Using the small cold chisel, knock out the cut piece, Figure 7–5. Now the solder bushing can be removed. What a performance just to replace a trap.

Figure 7–2 A solder bushing.
Courtesy Mueller Industries

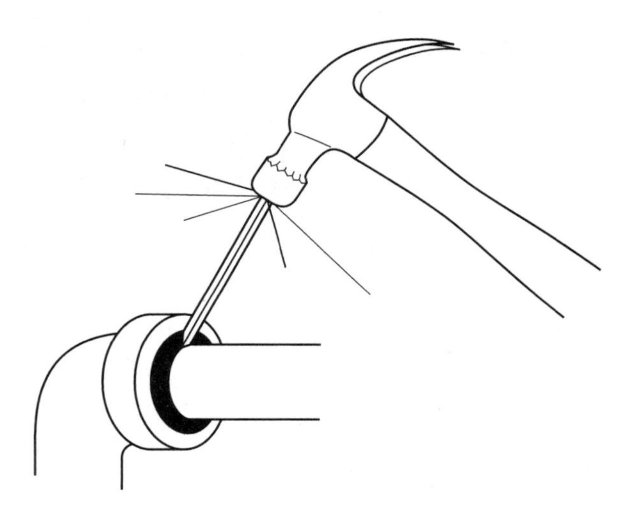

Figure 7–3 Cutting the trap arm from the solder bushing.

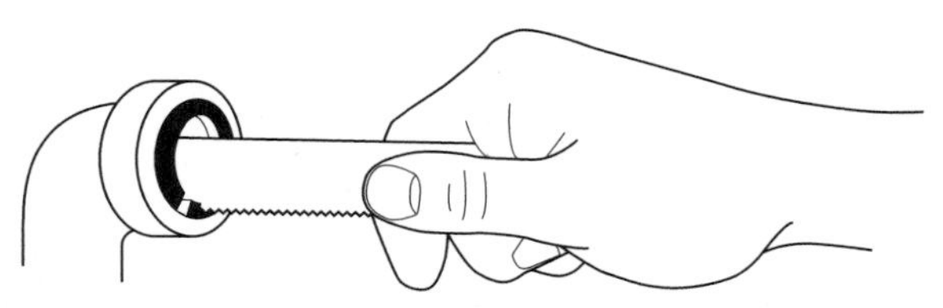

Figure 7–4 Making saw cuts in a solder bushing.

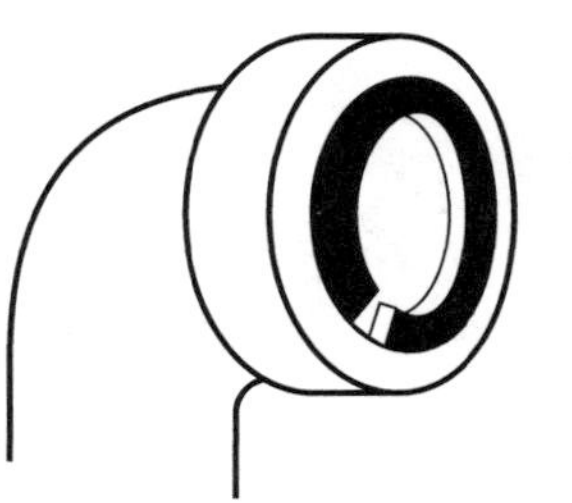

Figure 7–5 One-half-inch piece knocked out of solder bushing.

With the old trap and solder bushing removed, the trap can be replaced. With a female pipe thread (fitting) at the wall, a male adapter, Figure 7–1 (D), will be needed, and a PVC adapter can be used at about one-fourth the cost of a brass adapter. A chrome-plated brass P-trap could be used for aesthetic reasons if the trap is exposed. If the trap is concealed in a cabinet, a PVC trap can be used. When measuring pipe sizes, remember that plumbing pipe and fittings are measured by inside diameter (I.D.). Thus, a 1 1/2- × 1 1/4-inch-trap adapter would have a male thread measuring approximately 1 7/8 inches O.D. Before starting the adapter in the fitting, wrap one layer of Teflon® tape dope on the threads, beginning at the end of the fitting and wrapping clockwise toward the middle.

NOTE: Plumbing codes and regulations must be followed in areas where work is being done.

8

All About Faucets

TOOLS NEEDED

Basin wrench

6- or 8-inch adjustable wrenches

Large slip-joint pliers

Blade and Phillips screwdrivers

MATERIALS NEEDED

Faucets as selected

Braided flexible stainless steel supplies

Teflon® tape

Replace or Repair?

To replace or repair is a perplexing question. Let's consider replacing first. Is that old dripping faucet getting on your nerves? Have you wanted to replace it with a bright, shining new kitchen or bathroom faucet and then wondered if you could do it? I can tell you the answer. Yes, you can do it. Take Saturday morning off and visit your home improvement store. Browse through the faucets section of the plumbing department. You will have lots to choose from, from the new pull-out spray head to the antique-style of the 1920s.When you have made your selection, I'll help you find your way through the problems of removing the old faucet and installing the new one.

Removing the old faucet is the hardest part of the job. Just knowing how to go about it will make the job much easier. We'll go about it one step at a time.

There are so many brands and styles of faucets available to choose from that I am only going to show you two brands that have many styles available in their lines. If, as they say, imitation is the sincerest form of flattery, this is surely proven in any faucet department. The first step in either repair or replacement is to turn off the hot- and cold-water supply to the faucet. The control valves, Figure 8–1

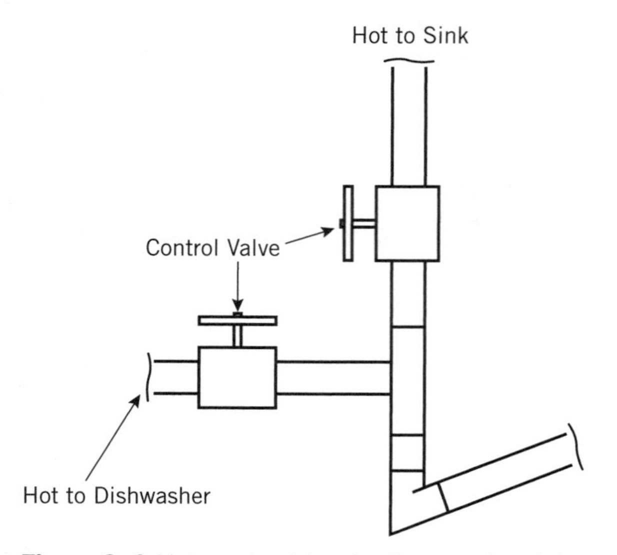

Figure 8–1 Hot- and cold-water lines under sink.

(compression stops), will be under the sink countertop. If these valves do not control the water, refer to Chapter 3.

Kitchen sink faucets are either top or bottom mounted as seen in Figure 8–2. Top-mounted faucets sit on top of the sink. Bottom-mounted faucets are inserted up through the mounting holes in the sink and secured to the sink by locknuts on the sink top.

The hardest part of removing and installing kitchen sink faucets is getting into the cabinet space to do the work. Almost everyone has a garbage disposer. That means a waste pipe connection from the disposer to the sink trap connection and the piping from the sink trap to the waste connection in the wall. Then there is electric wiring to the disposer and the drain connection from the dishwasher to the disposer to work around. And almost certainly a cold-water connection with piping to an ice

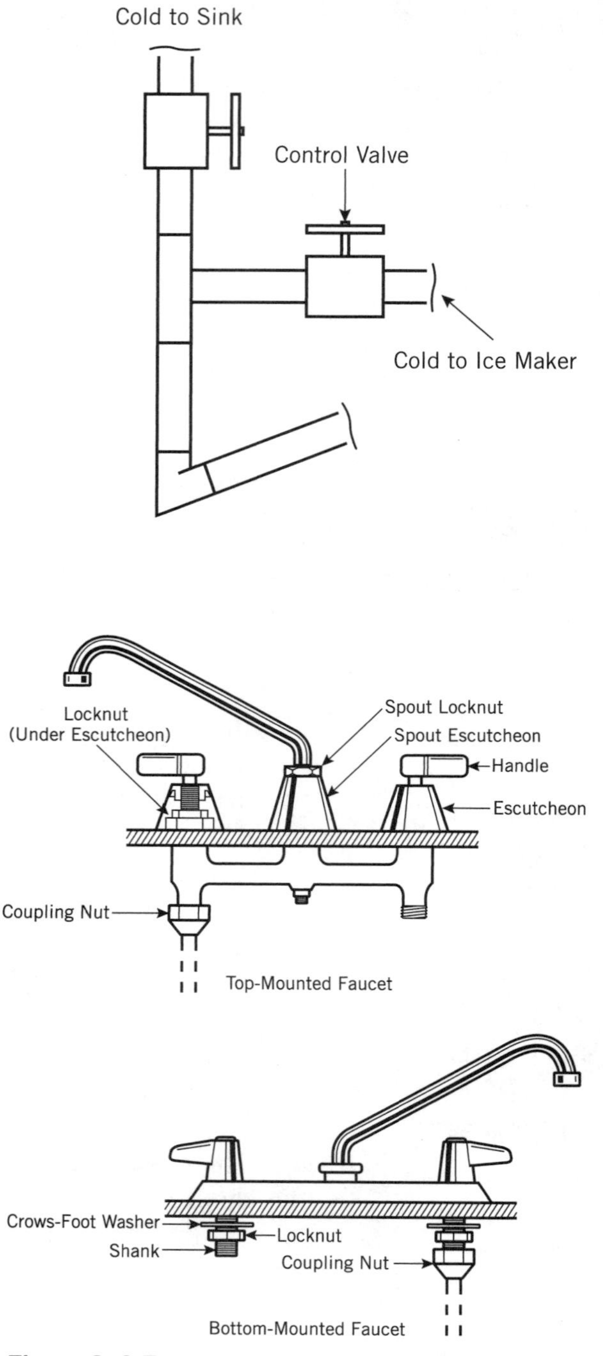

Figure 8–2 Top- and bottom-mount sink faucets.

maker in the refrigerator. You have to work around all these obstacles to remove an old faucet and install a new one. This is not an easy job, but I will try to make it as painless as possible. You can do it.

For a top-mounted faucet, when you are sure that both hot and cold water to the faucet are shut off, remove the handles and the escutcheons. Turn the spout locknut counterclockwise and remove it. Take the spout out and remove the spout escutcheon. Turn the locknut under the escutcheon counterclockwise, and remove it; the faucet will now drop out.

If the faucet is a bottom-mounted type, shown in Figure 8–2, and you have worked your way into a position under the sink and in the cabinet where you can get at the faucet connections, the next step is to use your basin wrench, Figure 2–6, to loosen and remove the coupling nuts connecting the water supply to the faucet. If the existing supply tubes are not braided stainless steel flexible supplies, discard them. We will use new braided flexible stainless steel supplies to connect the new faucet.

If the locknuts on this faucet are made of plastic the new basin wrench, Figure 2–7, makes removing them easy.

If the faucet is secured to the sink by threaded bolts and nuts, use your adjustable wrenches to remove the nuts. If the threads are rusted, WD-40 or Liquid Wrench penetrating oil may help in turning the nuts. If the bolts twist off, so be it. The faucet will come out.

If the faucet is secured by locknuts, use your basin wrench and turn the locknuts counterclockwise to loosen and remove them; the faucet can now be taken out.

If the faucet is a bottom-mount type, after the coupling nuts shown in Figure 8–2 are loosened and removed, turn the locknuts counterclockwise to loosen and remove them and the faucet can be lifted out of the sink. Bottom-mount faucets are often secured to the sink by plastic locknuts, and they can be very hard to loosen and remove. A basin wrench is almost indispensible for removing and installing faucets.

There is a new basin wrench that makes removing old plastic locknuts (and installing new ones) unbelievably easy. The notched ends of the plastic nut basin wrench, Figure 2–7, are designed to self-center on two-, three-, four-, and six-tab nuts and to fit metal hex nuts. Made of 16-gauge cold rolled steel, it features a one-piece 11-inch tubular steel body to reach faucet water supply shanks. This is a tool that should be in the tool box of every do-it-yourselfer. (It will also be in the professional plumber's tool box.)

If either faucet has a hose and spray, the hose connection to the faucet can be removed after the faucet is out of the sink. If you've gotten this far, take a break, you've earned it. Then we'll go on.

When you buy a new faucet, the instructions for installing it will be packed with the faucet. You can follow these instructions and not go wrong. I can, however, give you a tip or two that will make the installation much easier.

Braided flexible stainless steel supplies are now available for connecting the water supplies to the new faucet. Because these connectors are flexible, they do not have to be an exact length; just make sure they are a little longer than necessary. They can be shaped into an offset if needed to make the connection fit perfectly. A braided stainless steel connector for faucets is shown in Figure 8–3. If you haven't used these before, I can guarantee that once you use them, you will never use any other type of flexible connector again. Different brands of faucets will connect in different ways, but there are braided flexible connectors that will fit any brand of faucet.

Quite often it is possible and practical to connect the hose and spray, if one is used, to the faucet before the new faucet is mounted. In most installations it is also possible to connect the flexible supplies to the faucet shanks before the faucet is mounted. The fewer connections you have to make while lying on your side or back under the sink the better.

Types of Faucets

The following information about faucets will help you identify the faucet now in use and will also help in the selection of replacement faucets. Many of these faucets have looped handles for people suffering from arthritis.

Delta Faucets

The faucet shown in Figure 8–4 is the Delta Waterfall® series for four-hole installation. It is a single-lever faucet with a loop handle. The soap dispenser provides extra convenience.

The faucet shown in Figure 8–5 is the Delta Signature™ Series single-handle kitchen faucet with pull-out spray. The single handle is easy to operate and Americans with Disabilities Act (ADA) compliant.

Delta's Model 400 single-handle faucet with vegetable spray is shown in Figure 8–6. Its solid brass construction ensures long-lasting quality.

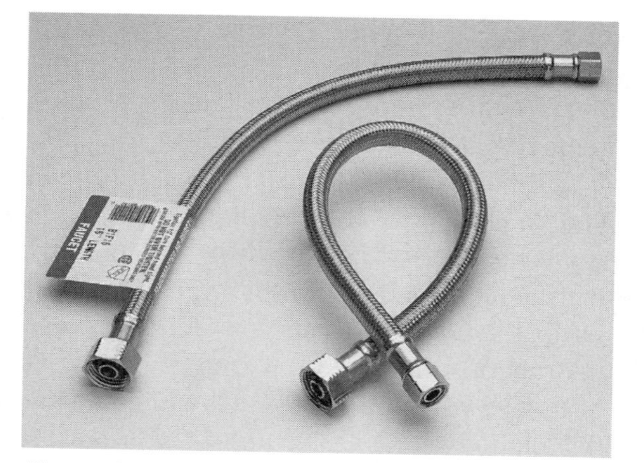

Figure 8–3 Braided stainless steel faucet connectors.
Courtesy Fluidmaster

Figure 8–4 Delta Waterfall® Series kitchen faucet.
Courtesy Delta Faucet Co.

Figure 8–5 Delta Signature™ Series pull-out spray kitchen faucet.
Courtesy Delta Faucet Co.

Figure 8–6 Delta single-handle kitchen faucet with spray.
Courtesy Delta Faucet Co.

Figure 8–7 Delta Innovations™ single-handle lavatory faucet with pop-up waste.
Courtesy Delta Faucet Co.

Figure 8–8 Delta C Spout two-handle lavatory faucet with pop-up waste.
Courtesy Delta Faucet Co.

Figure 8–9 Delta Neo Spout™ two-handle lavatory faucet with pop-up waste.
Courtesy Delta Faucet Co.

Figure 8–7 shows a Delta Innovations™ single-handle lavatory faucet with pop-up waste.

A Delta C Spout two-handle lavatory faucet with pop-up waste is shown in Figure 8–8.

A Delta Neo Spout™ two-handle lavatory faucet with pop-up waste is shown in Figure 8–9.

Price Pfister Faucets

Price Pfister's Genesis model kitchen faucet, Figure 8–10, is made with ceramic disc

Figure 8–10 Price Pfister Genesis single-lever kitchen faucet.
Courtesy Price Pfister

Figure 8–11 Price Pfister two-handle kitchen faucet.
Courtesy Price Pfister

Figure 8–12 Price Pfister Genesis kitchen faucet with vegetable spray.
Courtesy Price Pfister

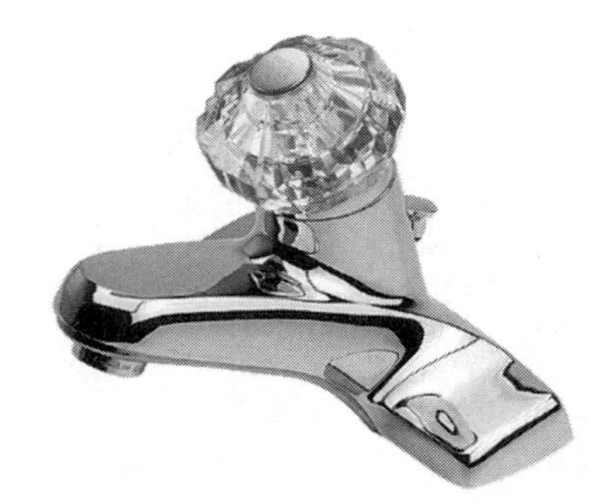

Figure 8–13 Price Pfister Genesis 4-inch center-set lavatory faucet with pop-up waste.
Courtesy Price Pfister

valves that are guaranteed never to leak. It is also available with loop or lever handles.

Price Pfister's Bedford series two-handle kitchen faucet with ceramic disc valves is available with or without vegetable spray.

Figure 8–11 shows Price Pfister's two-handle faucet.

The Price Pfister faucet shown in Figure 8–12 is a single-lever handle kitchen faucet with Teledyne in-spout filter and vegetable spray.

Price Pfister's 4-inch center-set lavatory faucet, Figure 8–13, has a pop-up waste option.

The Price Pfister Parisa single-handle lavatory faucet, shown in Figure 8–14, has a pop-up waste.

Figure 8–15 shows the Price Pfister Genesis 4-inch centerset lavatory faucet. It has a looped handle, making it easy to operate.

Sterling Plumbing Faucets

Sterling's Progression™ line single-lever kitchen faucet is shown in Figure 8–16. It has a chrome finish and vegetable spray.

The Sterling Tribute™ single-lever chrome finish kitchen faucet, shown in Figure 8–17, mounts in a single hole. It also has a vegetable spray.

Sterling's Eminence™ line two-handle lavatory faucet, Figure 8–18, has porcelain lever handles. It comes with a chrome pop-up waste.

Figure 8–14 Price Pfister Parisa single-handle lavatory faucet with pop-up waste.
Courtesy Price Pfister

Figure 8–15 Price Pfister Genesis 4-inch center-set lavatory faucet with looped handle and pop-up waste.
Courtesy Sterling Plumbing

Figure 8–16 Sterling Progression model single-lever kitchen faucet with a vegetable spray.
Courtesy Sterling Plumbing

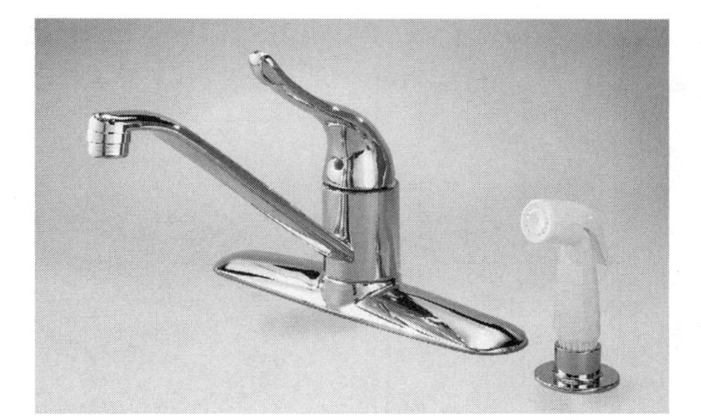

Figure 8–17 Sterling Tribute model single-lever kitchen faucet with vegetable spray.
Courtesy Sterling Plumbing

The Sterling Progression™ single-lever lavatory faucet, Figure 8–19, has a chrome finish and is complete with a pop-up waste.

Faucets with looped handles help people who have arthritis problems. Quite often it is difficult to identify the brand of an older faucet, especially an old washer-type faucet. The Appendix at the end of this book contains full-size photos of faucet repair parts, stems, O rings, and faucet seats, to help in identifying parts needed for repairing washer-type faucets.

Figure 8–18 Sterling Eminence model two-handle lavatory faucet.
Courtesy Sterling Plumbing

If you have a sink with four mounting holes and wish to install a faucet that does not have a spray hose, a chrome-plated device called a cock-hole cover will cover the fourth hole. The fourth hole can also be used to mount a liquid soap dispenser or a hot-water dispenser.

The Sterling Eminence model is a popular two-handle lavatory faucet.

The Sterling Progression model single-lever lavatory faucet is shown in Figure 8–19.

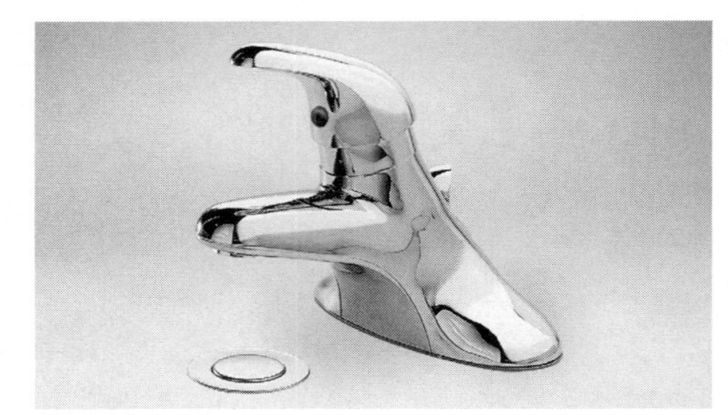

Figure 8–19 Sterling Progression model single-lever lavatory faucet.
Courtesy Sterling Plumbing

9

Protecting Your Family's Health

Toilet Tanks

There is only one thing besides the working parts of a toilet tank that belong in the tank: water just as it is delivered to your home by your municipal or private water system. There should be no bottles or plastic containers containing some miracle cleaner. The shelves of many stores contain products that claim to clean the bowl with every flush. Most, if not all, of these products contain strong chemicals. When these products are introduced into the toilet tank, the *water in the tank is contaminated.* Water from a contaminated toilet tank can end up in your drinking water supply.

Ballcocks—Fill Valves

If you have read Chapter 6, you know that I urge you to be very careful when you install a new fill valve for your toilet. If a ballcock or fill valve will permit back-siphonage to take place, polluted water in your toilet tank can be drawn into your drinking water supply. Back-siphonage is the flowing back of used contaminated or polluted water from a plumbing fixture or vessel into a water supply pipe due to a negative pressure in that pipe. How can this happen? Your domestic water supply depends on pressure to get from its source, whether it is a private well or a municipal water works, to your home. Anything that results in a serious drop in water pressure or a no-pressure condition is a direct threat to the health of everyone connected to the water supply. This usually happens because of a break in a water main. When a water main breaks, the main and affected areas go from a pressurized condition to a negative pressure condition. There are notices almost every day in the news to residents of affected areas: Boil water until further notice. So how does it directly affect you and your family? Let's set the scene for a serious accident: It is mid-morning, and no one is home. A backhoe operator working on a construction crew on the next street has broken a water main that supplies water to your home. The broken main is shut off, water from the main drains out, and a no-pressure condition exists in *all* the areas served by the main. Water in the piping in your home starts to flow back toward the main in the street. Someone had put a cartridge to clean the bowl in the toilet tank, and the water in the tank is heavily contaminated with chlorine. The ballcock in the tank is not an antisiphon type, thus allowing the contaminated water in the tank to siphon back into the cold-water piping. Shortly after noon the water main is repaired and water is turned back on. As the pressure is built back up, the contaminated water in the piping is spread throughout the house. At 3:00 p.m. school is finished for the day and your eighth grader comes rushing into the home, runs for the sink, fills a glass with water, and starts to drink. She swallows a small amount before the chlorine smell alerts her to the danger. Her mouth and esophagus are burned but she recovers.

This particular story is fiction, *but it can happen*. Back-siphonage from the toilet tank can be prevented by using an antisiphon ballcock or fill valve in the toilet tank. When purchasing a ballcock or fill valve, look for a label on the package stating that the unit is an antisiphon type.

A toilet bowl is made with an integral trap. The purpose of the trap is to provide a water seal that will prevent sewer gas from entering the room through the toilet bowl. When the flush lever is depressed, water from the tank enters the bowl, the bowl is emptied, and the tank ball drops into place, preventing water from the tank from refilling the bowl. Every ballcock or fill valve has a plastic or copper *refill tube*. As water is filling the toilet tank, the refill tube on the ballcock or fill valve is refilling the toilet bowl.

For bacteria in waste water to enter your drinking water system, back-siphoning must occur. This can be prevented by using an antisiphon fill valve or ballcock in the toilet tank. Some antisiphon ballcocks can be identified by a series of holes around the top area, either above or below the valve housing. The holes let air in, breaking a vacuum and preventing siphoning. Other antisiphon units may not have readily visible holes but do have a large air entrance around the valve operating lever for this purpose. An antisiphon fill valve is shown in Figure 6–2.

Garden Hoses

Another common offender in causing contamination of drinking water supplies is the ordinary garden hose. Garden hoses are attached to sillcocks for outside watering purposes and are also often connected to sprayers used for applying fertilizers, weed killers, and pesticides. If, as mentioned earlier, a negative-pressure condition should exist while a sprayer is attached to a hose, contaminants could then be siphoned back into the house drinking water piping. The correct type of vacuum breaker used in conjunction with a garden hose can prevent back-siphonage.

The following are examples of how a garden hose can cause contamination.

1. A garden hose is being used to trickle feed a garden or soak a lawn, with the hose end lying in a pool of water. For any of the reasons mentioned earlier, a negative-pressure condition occurs, and groundwater containing fertilizers or chemicals is siphoned back into the house piping system through the sillcock. The vacuum breaker shown in Figure 9–1 is designed for use on an angle sillcock. The vacuum breaker shown in Figure 9–2 is designed for use on freezeproof sillcocks.

2. A hose end sprayer is being used to apply fertilizer, pesticides, or weed killer as shown in Figure 9–3. For some reason the person doing the spraying is called away and sets the sprayer down. The sprayer may lay there for hours. If a negative-pressure condition occurs during the person's absence, the contents of the sprayer could be siphoned back into the house's water piping system.

3. The hose is being used to fill a small swimming pool used by children (or it

Figure 9–1 An angle sillcock type vacuum breaker.
Courtesy Watts Regulator

Figure 9–2 This vacuum breaker is designed for use on freezeproof sill cocks.
Courtesy Watts Regulator

Figure 9–3 A sillcock vacuum breaker will prevent back-siphonage when hose end sprayers are used.
Courtesy Watts Regulator

could be any pool or tank) and a negative-pressure condition occurs. Water from the pool or tank could then be siphoned back into the house's water piping system.

These are only a few examples of contaminants being back-siphoned through garden hoses into the house water piping system. Plumbing codes and regulations now require the use of vacuum breakers on all types of sillcocks. These vacuum breakers are available at home improvement and hardware stores.

Here is another example; this actually happened! The water services to 75 apartments housing approximately 300 people were contaminated with chlordane and heptachlor in a Pennsylvania city in 1980. The insecticides entered the water supply system while an exterminating company was applying them as a preventive measure against termites.

The pesticide contractor was mixing the chemicals with water in a tank truck, *using a garden hose connected to a sillcock in one of the apartments.* The end of the hose was submerged in the tank containing the pesticides. At the same time a workman had shut off the water main supplying the apartments and had cut the main, preparing to install a 6-inch valve on the main. When water started draining from the cut main, a back-siphonage condition was set up. As a result, the mixed chemicals were back-siphoned out of the truck through the garden hose and into the water system, contaminating the 75 apartments.

Repeated efforts to clean and flush the piping were not satisfactory, and a decision finally was made to replace the water lines and all the plumbing that was affected. Residents were told not to use the tap water for any purpose, and water was trucked into the area by volunteer fire departments. Residents were without their normal water supply for 27 days. All the expense and inconvenience could have been avoided if a hose vacuum breaker had been installed on the sillcock to which the garden hose was attached.

Lawn Sprinkling Systems

Many lawn sprinkling systems are designed using subsurface types of sprinkler heads. The heads pop up above ground level when the system is activated and retract when the system is not in use. The soil around the heads contains naturally occurring contaminants as well as bird and animal droppings and residue of fertilizers, pesticides, and weed killers. If a negative-pressure condition occurs, back-siphonage can draw contaminated ground water into the sprinkler piping system. If the sprinkler piping is connected to the potable water piping system, the entire system will be contaminated. This can be prevented by using a vacuum breaker of the type shown in Figure 9–4.

When the water supply valve upstream of the vacuum breaker is open and a negative pressure is created in the supply line, the disc float drops, opening the atmospheric vent and at the same time closing the orifice opening. This prevents the creation of a vacuum in the discharge line downstream of the vacuum breaker and prevents back-siphonage. The cutaway view of this vacuum breaker, Figure 9–5, shows how an

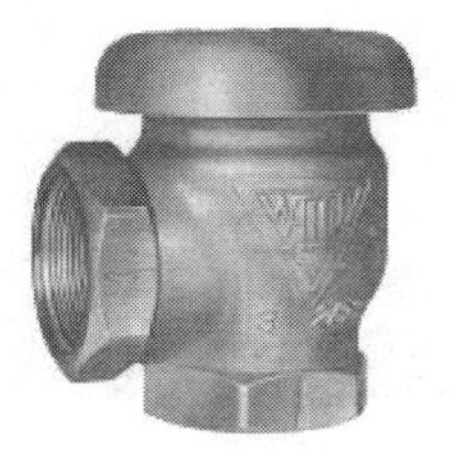

Figure 9–4 An atmospheric-type vacuum breaker.
Courtesy Watts Regulator

atmospheric-type vacuum breaker works to prevent back-siphonage.

When this type vacuum breaker is used, to be effective it should be installed as shown in Figure 9–6. The figure shows the correct way to install an atmospheric-type vacuum breaker when the sprinkler system is connected to the potable water system.

Plumbing codes and regulations require the installation of backflow preventers on the downstream (building) side of water meters. This prevents contamination of

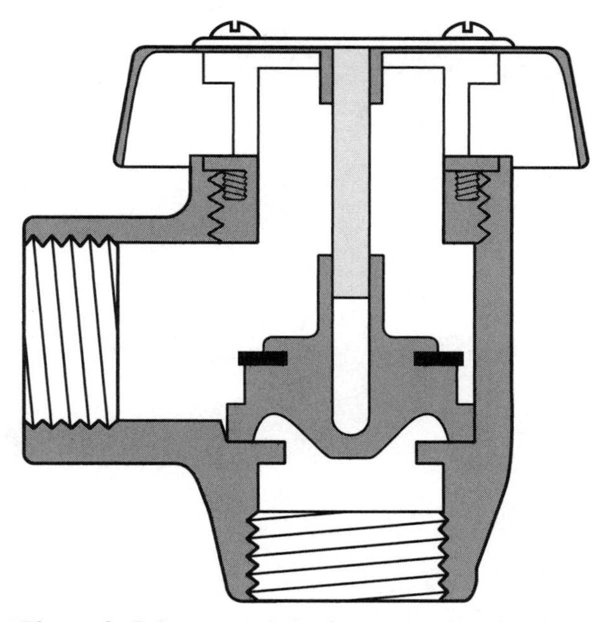

Figure 9–5 Cutaway view of an atmospheric-type vacuum breaker.
Courtesy Watts Regulator

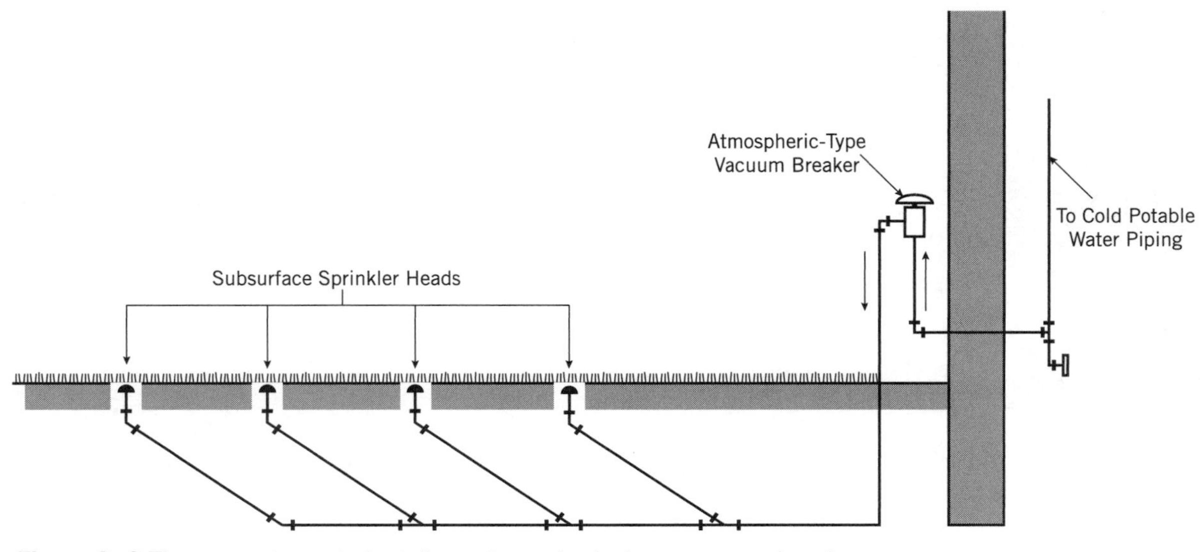

Figure 9–6 The correct way to install an atmospheric-type vacuum breaker.
Courtesy Watts Regulator

public water supply systems by preventing contamination from an individual unit from entering water mains. One type of residential dual-check valve backflow preventer is shown in Figure 9–7.

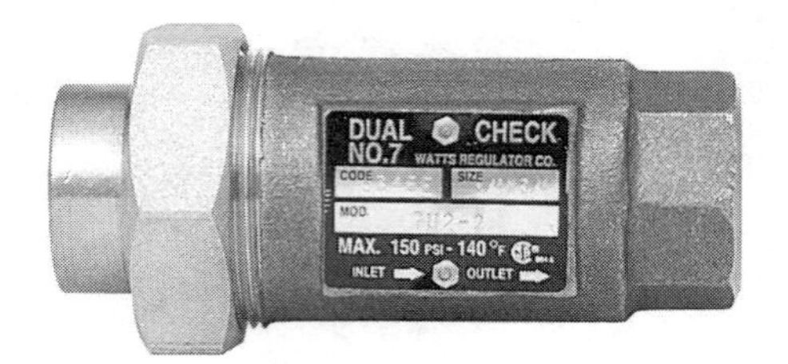

Figure 9–7 A dual-check backflow preventer.
Courtesy Watts Regulator

10

Removing an Old Pop-Up Waste

TOOLS NEEDED

Hacksaw blade

Slip-joint pliers

This is a short chapter, but it is a very important one. You've bought a new lavatory faucet with a pop-up waste. Before you can install that new pop-up waste that came with the new lavatory faucet, you must remove the old pop-up waste. This can be as easy as pie if the old waste is the type shown in Figure 10–1. Turn the packing nut clockwise (looking at it from the front) to loosen and remove the nut. Remove the lever ball assembly from the pop-up body. Then lift out the stopper.

The pop-up is secured in the lavatory by the locknut. Use your large slip-nut pliers to turn this locknut counterclockwise (looking at it from below) to loosen and remove it.

Push up on the pop-up body, and slide the flat washer and rubber gasket down and off of the pop-up body. The old pop-up body can now be lifted out of the lavatory.

If the pop-up waste is the type shown in Figure 10–2, it can be difficult. The chrome-plated flange inside the lavatory is screwed onto the pop-up body. Turn the packing nut clockwise (looking at it from the front) to loosen and remove the nut. Remove the lever ball assembly from the pop-up body. Then lift out the stopper.

The pop-up is secured in the lavatory by the locknut. Use your large slip-nut pliers to turn this locknut counterclockwise (looking at it from below) to loosen and remove

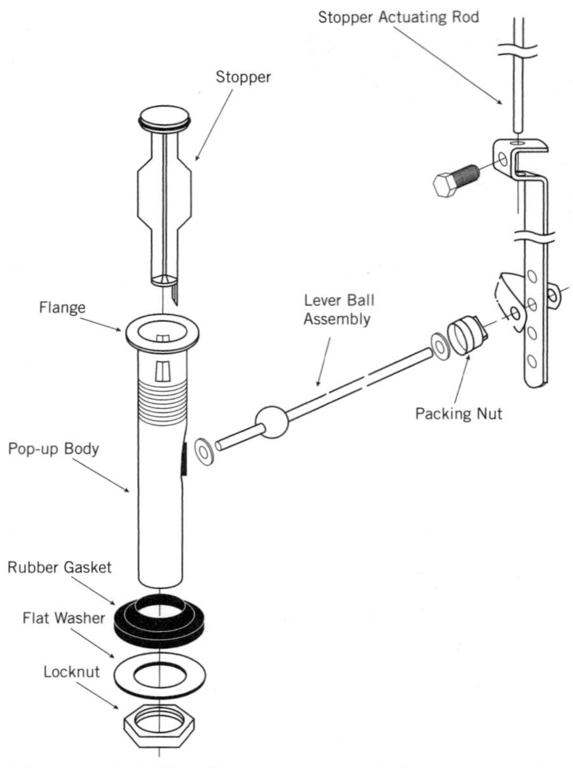

Figure 10-1 The flange is part of the body of this pop-up waste.

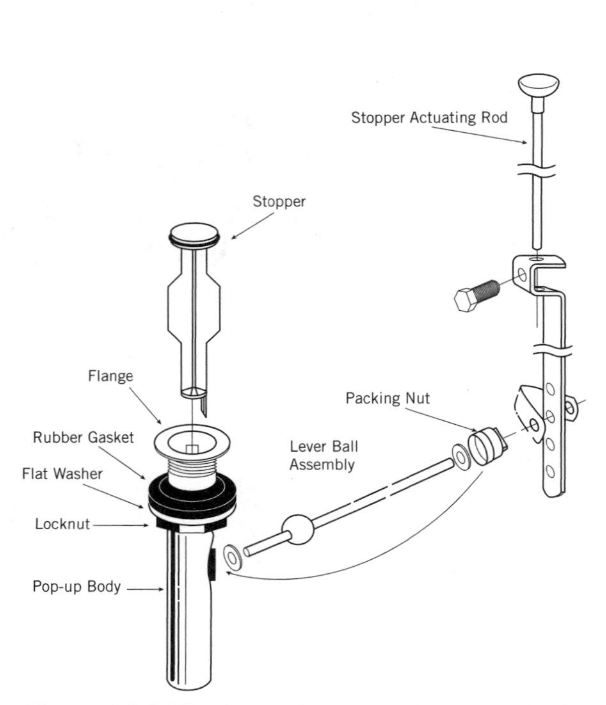

Figure 10-2 The flange is screwed onto the body of this pop-up waste.

it. Grasping the pop-up body, open your slip-joint pliers and try to turn the flange counterclockwise. If you're lucky, the flange can be unscrewed from the pop-up body. If you're not lucky, we'll do it the hard way. Tighten the locknut to secure the pop-up in the lavatory. Wrap a layer of duct tape around a fine-toothed hacksaw blade as shown in Figure 10–3 and saw almost, but not quite, through the flange. Do not mar the porcelain of the lavatory.

Loosen the locknut. Push up on the pop-up body, and while holding it with one hand, use your slip-joint pliers to grasp the rim of the

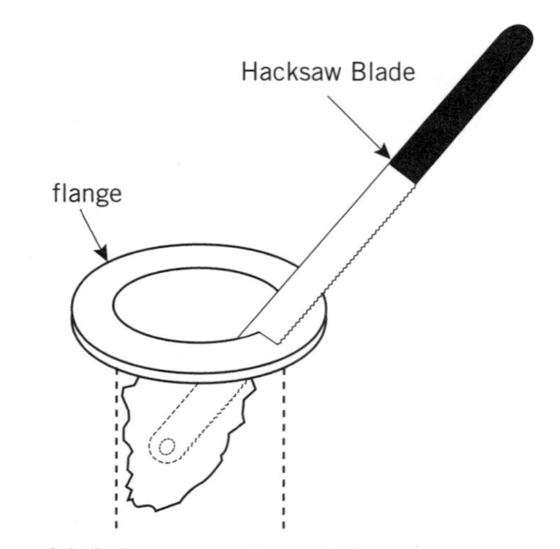

Figure 10-3 Removing the old flange.

flange and break the rim off. Now the pop-up body can be removed from the lavatory.

Now you can get on with the job of installing the new faucet and pop-up waste. When you are installing the new pop-up waste assembly, be sure to put a ring of plumbers putty between the flange and the lavatory before you tighten the locknut to secure the flange. Putty must be used between the flange and the lavatory to prevent water from leaking out.

11

Repairing a Trip-Lever Bath Drain

TOOLS NEEDED

Screwdrivers

Long-nose pliers

Liquid Wrench or WD-40

The trip-lever bath drain shown in Figure 11–1 is subject to two common problems. When the raised portion, or cam (over which the lever must ride), on the back of the faceplate becomes worn, the lever will no longer stay in a set position. Also, if the lever is not used and is allowed to stay in one position, soap, grime, hair, and so on, present in the bath water will build up around the brass plug, making it immovable.

This is one plumbing part that can be replaced easily. Every hardware and home improvement store carries replacements. There are two replacement types. One type has a lever that moves up and down, lifting the brass plug off the seat on the downward motion, allowing the water to run out, and setting the plug back on the seat on the upward motion, holding the water in the tub. The other type has a lever that moves 180° from right to left, in an up and over motion.

Removing the Faceplate

To replace the faceplate, the linkage and plug attached to it must be removed through the overflow piping. Turn the two screws that secure the faceplate in a counterclockwise direction to loosen and remove them. Pull the faceplate out

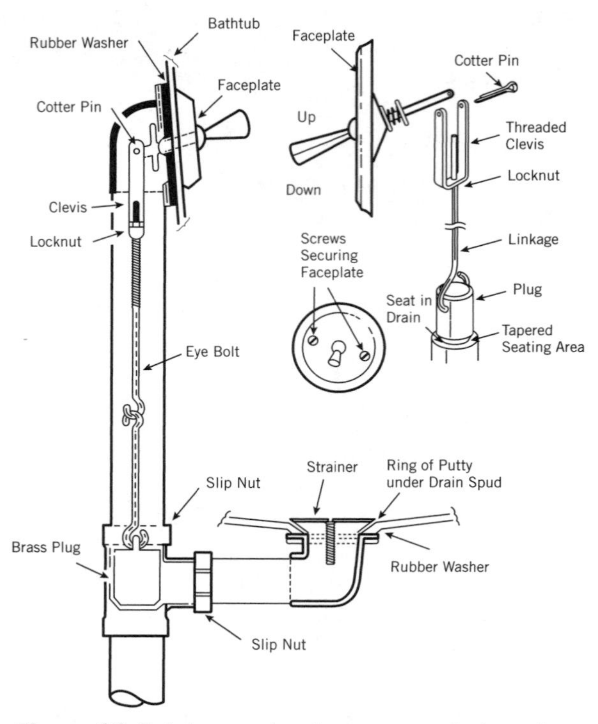

Figure 11-1 A brass-plug type connected waste and overflow fitting.

and up; the linkage and the plug should come up and out through the overflow opening. *Do not force it.* If the linkage and plug do not come out easily, *stop* and spray a penetrating oil (Liquid Wrench or WD-40, for example) into the overflow opening. Go read the funnies or listen to the news on TV to give the penetrating oil time to work. The oil will run down the overflow pipe and will usually free the plug in a short time. Be patient; you may have to spray two or three times over several hours to allow the oil to do its work.

When the plug and linkage have been removed, use long-nose pliers to remove the cotter pin securing the faceplate to the clevis. Discard the old faceplate and connect the new one to the linkage using the old cotter pin. Clean the outside surface of the brass plug, using fine sandpaper or steel wool, and then grease the plug lightly before inserting it and the linkage into the overflow piping. I have found that Plumbers Heat-Pruf Grease is the best all-around lubricant for plumbing repair jobs. Secure the faceplate to the overflow piping and test the trip lever for proper operation. As shown in Figure 11–1, when the lever is in the down position, the brass plug is lifted off the drain seat. When the lever is in the up position, the plug should be positioned on the drain seat, thus holding the water in the tub.

Adjusting the Linkage

If the linkage was adjusted correctly before replacing the faceplate, no additional adjustment should be necessary. Push the lever to the up position; if it will not go completely up, the linkage rod must be shortened. Remove the faceplate and lift out the rod and plug. Loosen the locknut and turn the linkage rod into the clevis three or four turns, tighten the locknut, and repeat the test procedure. When the plug is seated correctly, the lever can be set in the up position without meeting resistance. When the lever is in the up position but will not hold water in the tub, the rod must be lengthened by turning it out of the clevis. It may be necessary to repeat this process several times until the linkage is adjusted correctly.

The rocker-arm-type trip-lever drain shown in Figure 11–2 is also adjusted by turning the linkage rod into or out of the clevis and securing it with the locknut. The linkage and spring can be removed by removing the faceplate screws and pulling the faceplate up and out.

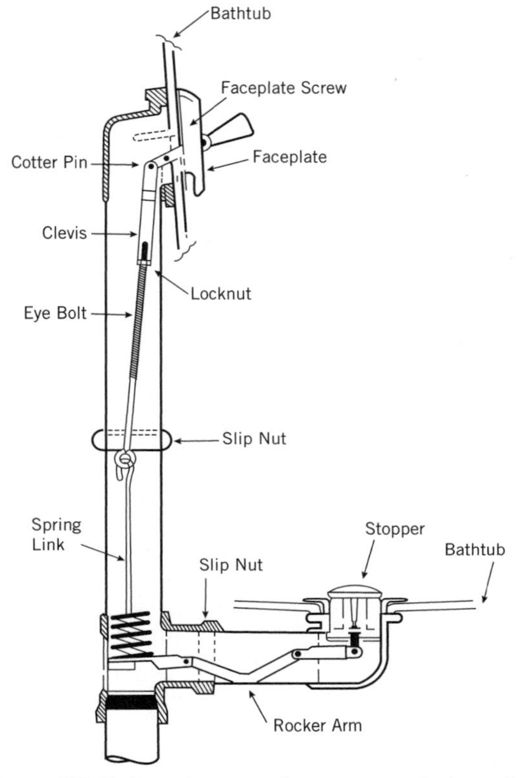

Figure 11-2 A rocker-arm type connected waste and overflow fitting.

Adjusting the Stopper

The stopper can be adjusted up or down by turning it clockwise or counterclockwise. The rocker arm may become corroded and often fails to operate properly due to hair, soap, and so on, that collect at the spring link. The remedy for this is to remove the linkage and spring and clean them. The rocker arm and stopper can be pulled up and out of the tub drain opening when cleaning or adjustment is necessary.

Spray the pivot points on the arm and the threads on the stopper with WD-40 or Liquid Wrench before putting the rocker arm back in place.

12

Installing or Resetting a Toilet

TOOLS NEEDED

Slip-joint pliers

Blade-type screwdriver

6- or 8-inch adjustable wrench

Hacksaw with fine-toothed blade

Plunger

Large sponge

MATERIALS NEEDED

Tank-to-bowl bolts, nuts, washers

Two closet bolts, nuts, washers

One wax ring

A remodeling project, installing a new tile floor in the bathroom, may lead to the purchase and installation of a new toilet. The big question in recent years has been whether or not the water-saver toilets really save water. When the federal government mandated that only toilets using 1.6 gallons of water per flush be used, a common complaint was that the toilets would not flush properly using only this amount of water. Having to flush toilets several times to clear the bowl certainly does not save water. When low-volume flush toilets first came out, there were problems. But, since then I have contacted several manufacturers, and they all told me they knew they had problems and the problems are gone now. The problems existed with the flapper-type tank balls. If you are replacing the flapper tank ball in a 1.6 gallon toilet tank, be sure to get the flapper tank ball made for that particular toilet.

The first step in installing a new tile floor in the bathroom is taking up the toilet. The control valve for water to the toilet is below the bottom of the tank on the left side. Turn the control valve handle clockwise to turn the water off. If this handle has not been turned for a long time, you may need to use pliers to turn the handle. If the handle is oval shaped, it is probably a die cast (pot metal) handle; care must be taken when trying to turn it. Lift the lid from the tank and flush the toilet. If water continues to flow into the tank, try turning the handle again. If water continues to flow, follow the instructions in Chapter 3.

Taking the Toilet Apart

With the water shut off, sponge out the water remaining in the tank. Then, looking at the water supply connections to the ballcock or fill valve, Figure 12–1, from below, turn the coupling nut "A" counterclockwise to disconnect it from the ballcock.

The tank is mounted on the bowl and secured by two bolts, Figure 12–2. Looking at the nuts from below, turn them counterclockwise to loosen and remove them. The tank can now be lifted up and off of the bowl.

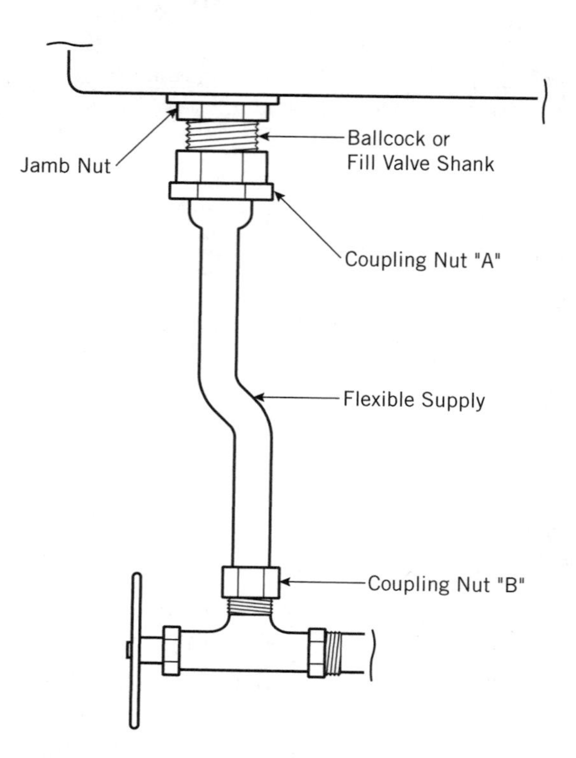

Figure 12-1 Turning the water off at the tank.

Removing the Toilet Bowl

Use the plunger to force most of the water out of the toilet bowl and then a sponge to get the rest of it out. If china (or plastic) bolt caps cover the closet bolts, Figure 12–3, remove the caps. Turn the nuts counterclockwise to loosen and remove them. Grasp the back of the bowl and gently pull up on the bowl. If the bowl adheres tightly to the floor, it can be damaged by rough treatment. When the bowl is loose, lift it out. Slide the old closet bolts out of the closet flange and discard the bolts.

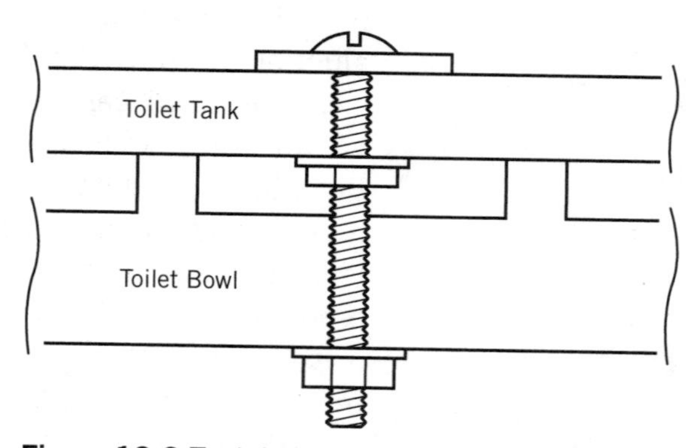

Figure 12-2 Tank bolts secure the tank to the bowl.

Resetting the Toilet Bowl

If you are just resetting the bowl or installing a new bowl, skip the instructions about raising the closet flange. If the bathroom has a wood floor, use woodscrews inserted through the round holes in the flange to secure the flange to the floor. The flange will only need to be raised if a ceramic tile floor has been added.

If you have had a ceramic tile floor installed, you have raised the floor level

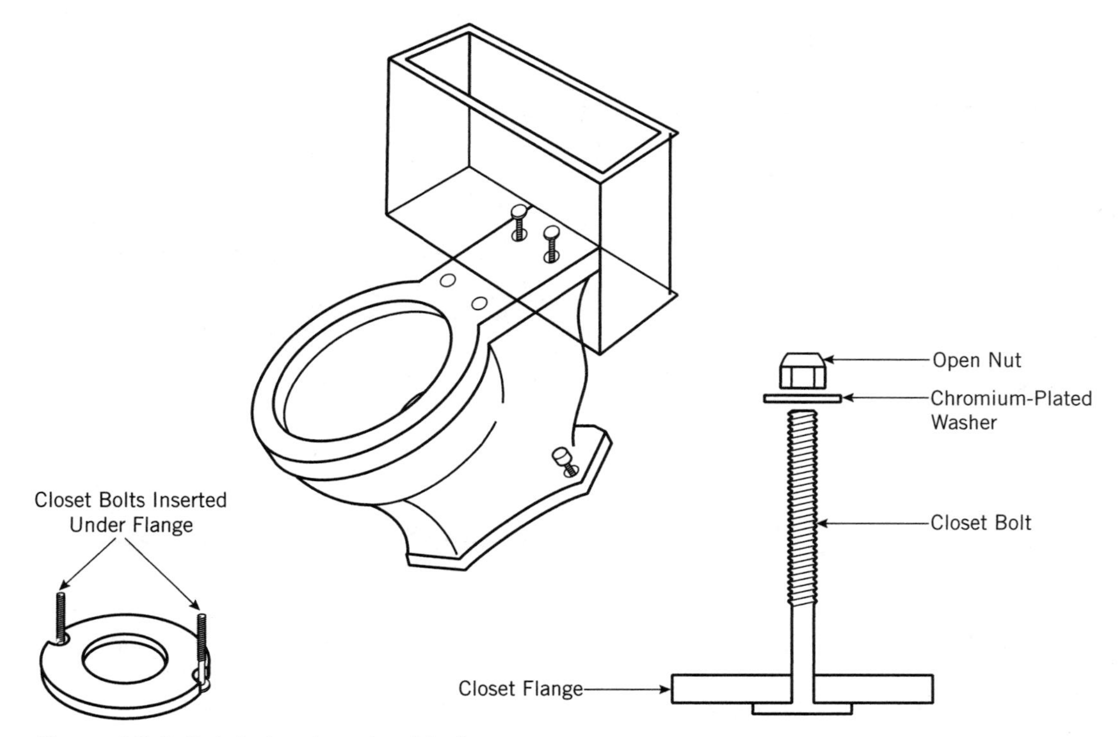

Figure 12-3 Detail showing closet bolts.

approximately 3/8 inch. The *top* of the existing closet flange is at or below the top of the tile, Figure 12–4, top.

If the top of the flange is at the floor level, the bowl will not be sealed correctly. There is a way out of this problem. Buy a PVC closet flange and, using a hacksaw, cut the bottom off the flange, leaving only the ring, Figure 12–4. Place the ring on top of the existing flange, lining up the holes and the slots in the ring. The top of the new flange will be at the correct height for setting the toilet bowl, Figure 12–4, bottom.

When resetting a toilet bowl or installing a new one, new closet bolts, nuts, and washers should be used. Closet bolts are made in two sizes, 1/4 and 5/16 inch. Use the 5/16-inch size. Set the closet bowl on edge and insert the wax ring on the horn as shown in Figure 12–5. Then with one hand on the front rim and the other on the back of the bowl, set the bowl carefully on the flange. The closet bolts will protrude through the mounting holes in the bowl. Push down on the bowl to set the wax ring. Drop the chrome-plated washers on the bolts and start the open nuts on the bolts. Screw the nuts down hand tight, rocking the bowl from side to side as you tighten the bolt. Then use a small adjustable wrench to tighten the nuts until they are fairly tight. We'll finish tightening them later.

If you are installing a new toilet, you will find new tank-to-bowl bolts and a new

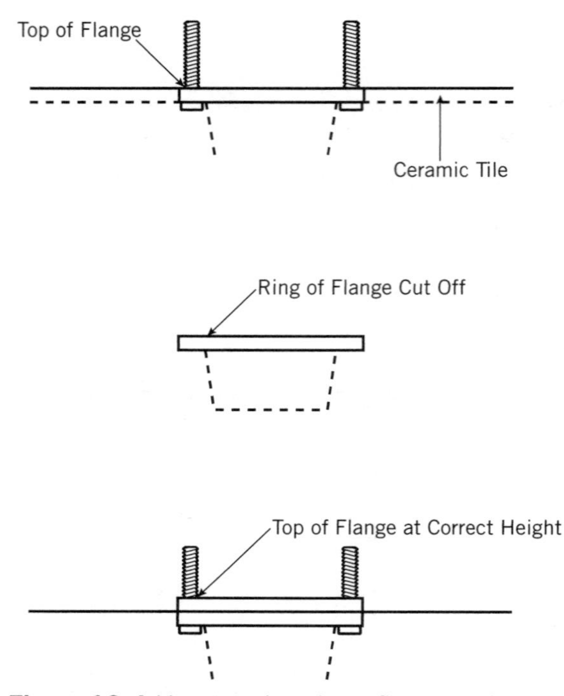

Figure 12-4 How to raise closet flange to the correct height.

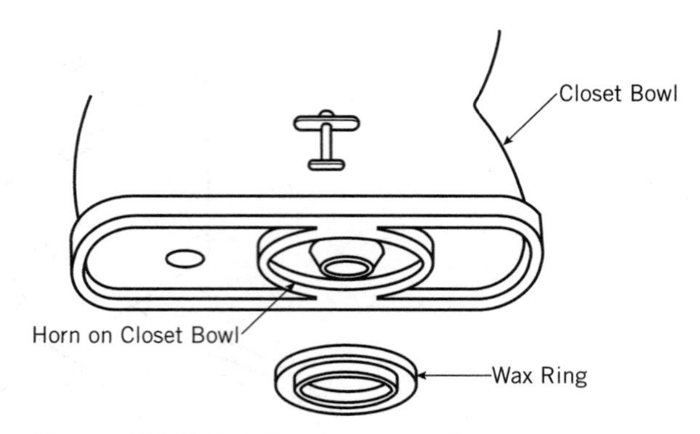

Figure 12-5 How to place wax ring.

tank-to-bowl gasket packed with the tank. If you are reinstalling the original toilet, you will need new tank-to-bowl bolts and a new tank-to-bowl gasket. This is not a problem; your home improvement store will have these parts. Set the tank on its side, slide the black rubber washers onto the bolts, and insert the bolts into the holes in the tank. Slide a flat washer on each bolt and start the nuts on each bolt. Use an adjustable wrench and tighten both nuts. Place the tank-to-bowl gasket over the flush valve and set the tank *gently* on the bowl, Figure 12–6. Sit down on the bowl and slide a flat steel washer on one of the tank bolts and start a nut on the thread. Repeat this on the other bolt. Now we get to the tricky part. Tighten the nuts evenly, going back and forth from one to the other, rocking the tank from front to back as you go and visually checking to see that the tank is level. Some closet bowls have raised places where the tank will touch when the bolts are tight. When you feel the tank touch these places, *stop* tightening the nuts. The tank and bowl are vitreous china, porcelain, and easily broken.

Now we'll go back to the closet bolts that secure the toilet to the floor. Visually check to make sure the toilet is positioned correctly and tighten the open closet nuts, going back and forth from one to the other. When they feel reasonably tight and the bowl seems firmly anchored, *stop*. Wait a day or two and then check the open nuts again. If bolt caps are to be placed over the bolts, saw off the threads above the nuts using a hacksaw with a fine-toothed blade. If sawing off the threads loosens the nuts, retighten them.

Connecting the Water Supply

Now that we're ready to connect the water supply, the job's almost done. One of the best things that's happened to the plumbing business in the last few years is the invention of braided flexible stainless steel supplies, Figure 5-6. Your home improvement store has these to fit almost every plumbing fixture and appliance. Just ask for one labeled Toilet. Measure the old supply and get one at least the same length; if it is a little longer, it will work perfectly.

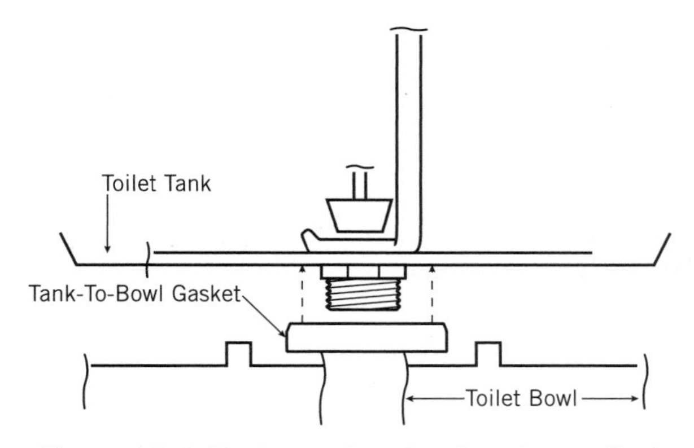

Figure 12-6 Placing tank-to-bowl gasket on flush valve.

Correcting Leaks at the Floor Line

Toilet bowls can leak at the floor line for several reasons: (1) The closet flange is not secured to a wooden floor. Plastic shims, available at hardware stores, may cure this problem. Secure the flange to the floor with flat head wood screws inserted through the round holes in the flange. (2) The closet flange is set too low. If the flange is not set at the correct height, it can be raised by cutting off the bottom of a PVC flange, as mentioned earlier. I have seen two wax rings used in place of one in an attempt to cure this problem. This usually results in wax being squeezed out from under the bowl, resulting eventually in leaks. (3) The closet bolts are not tight. Tighten the hex nuts on the bolts. If the leak still persists, take up the toilet and install a new wax ring. I prefer a wax ring that has a plastic horn that extends down into the closet flange.

13

PVC and CPVC Pipe and Fittings

PVC-DWV

How times have changed. It just dawned on me this morning that I can't remember the last time I saw a Roto-Rooter truck. Today almost all soil, waste, and vent piping, including the building sewer, is PVC-DWV. And PVC-DWV is so smooth inside that drains just don't stop up when it is used. For years plumbing codes in many areas required the use of cast-iron soil pipe for underground soil (human waste) and drainage piping. These codes also required the use of galvanized steel pipe for branch waste and vent pipe; in some areas lead pipe was required for waste connections to sinks, lavatories, and toilets.

Later, many of these areas permitted the use of DWV-grade copper tube for aboveground soil, waste, and vent piping. DWV copper is vulnerable to acids in human waste, and many areas and mechanical engineers now prohibit its use for that purpose.

The rising costs of labor and the comparatively low cost of PVC pipe and fittings have brought about the extensive use of PVC pipe and fittings in the plumbing industry. Here are some of the reasons for using PVC pipe and fittings:

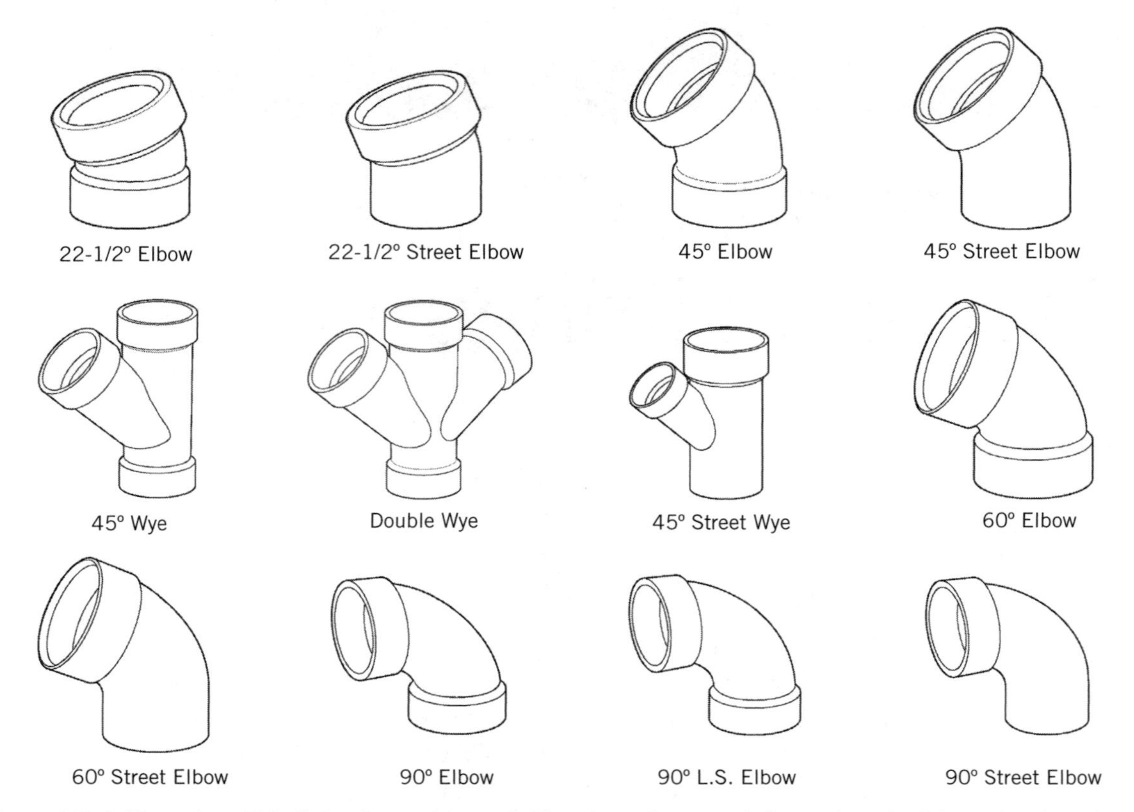

Figure 13–1 There is a PVC fitting for every need. Knowing what to ask for makes the job easier. (continued)
Courtesy Mueller Industries

1. No special tools are needed for installing PVC. The only tools needed are a rule to take measurements, a saw or cutter to cut the pipe, and a knife or file to ream out the burrs at the cuts. The extremely smooth inside surface is corrosion resistant, preventing the buildup of scale, rust, and foreign material that often impedes flow through metallic pipe.
2. PVC pipe and fittings are lightweight and are easily and quickly installed with chemically welded joints using solvent cementing.
3. PVC can be connected to existing cast-iron soil pipe, using either a poured lead joint or a threaded adapter.
4. PVC is virtually acidproof to any chemical used in recommended strengths around the home. Certain chemicals such as methyl-ethyl-ketone, used in paint removers and paint brush cleaners, should not be poured in PVC or any other drainage pipe.
5. PVC is rootproof. Roots cannot get into a building drain or building sewer if the joints are made properly.

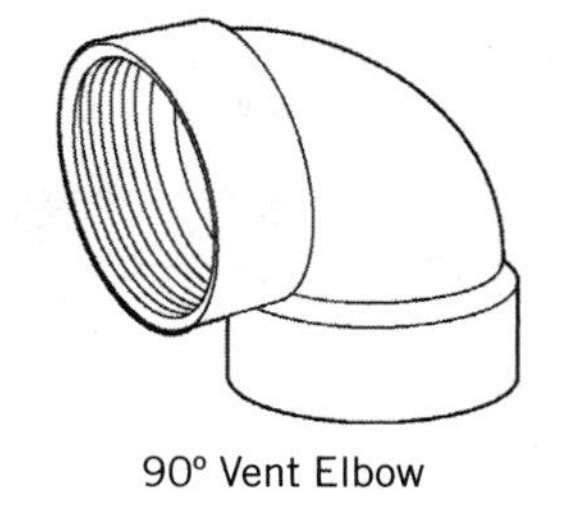

90° Vent Elbow

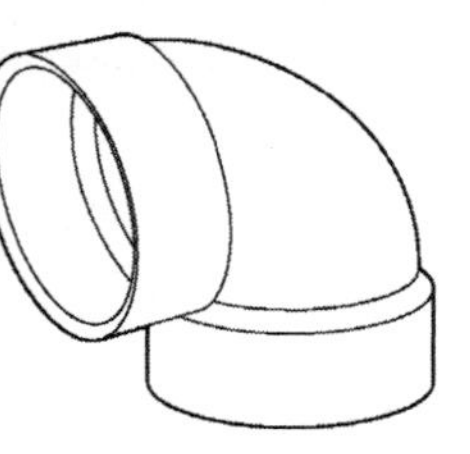

90° Vent Elbow

90° Vent Street Elbow

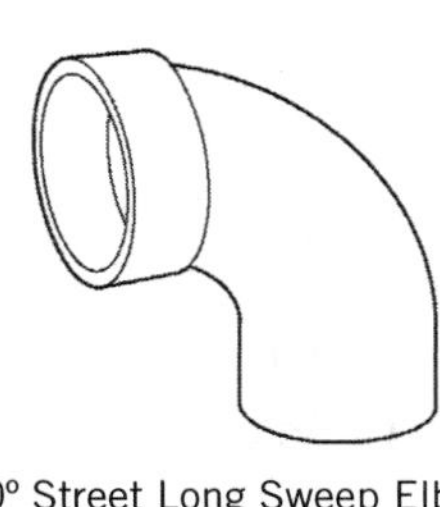

90° Street Long Sweep Elbow

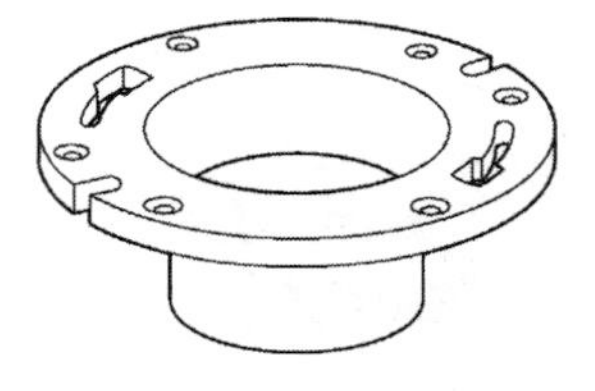

Hub End Closet Flange

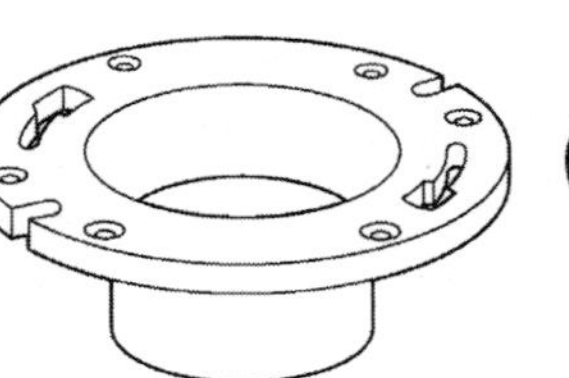

Spigot End Closet Flange

Offset Closet Flange

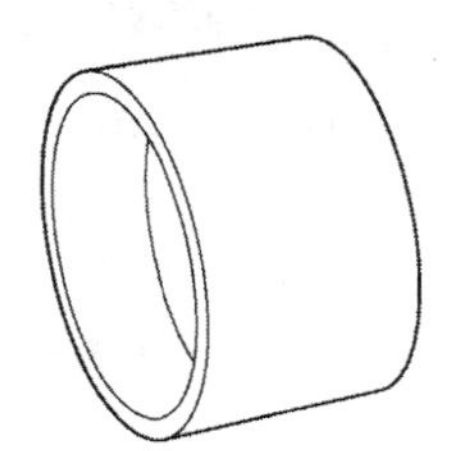

Coupling

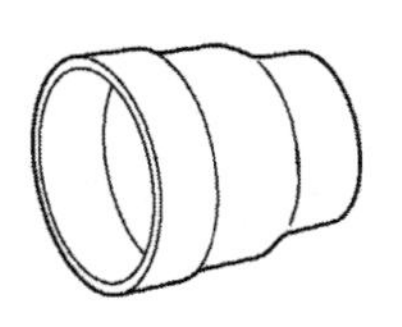

Reducing Coupling

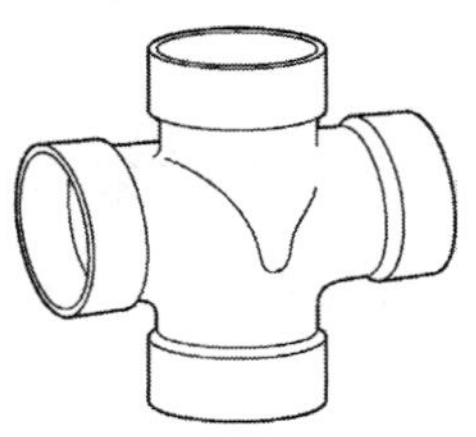

Sanitary Cross

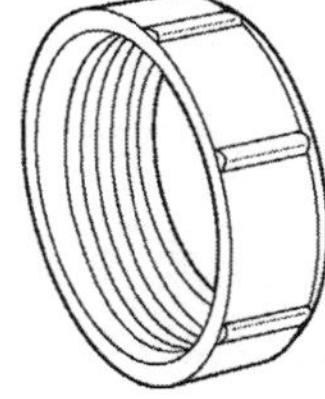

Plastic Slip Nut

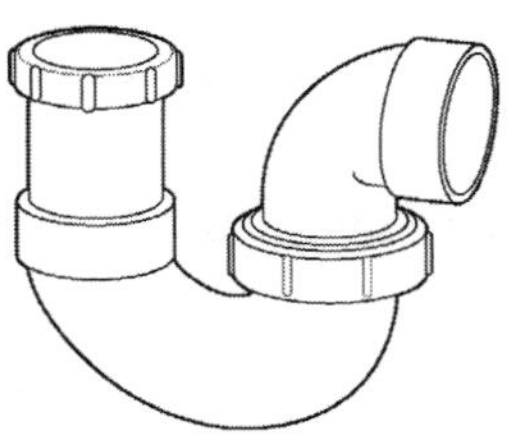

L.A. Pattern P-Trap

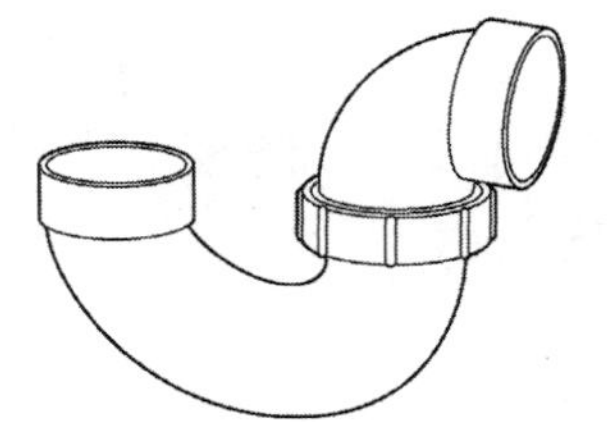

P-Trap with Union

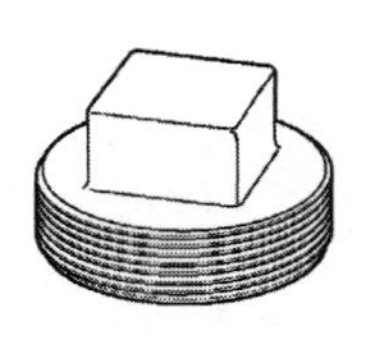

Thread Plug

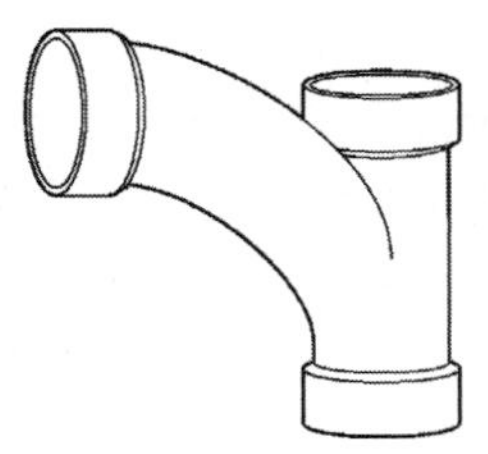

Long Turn Tee Wye

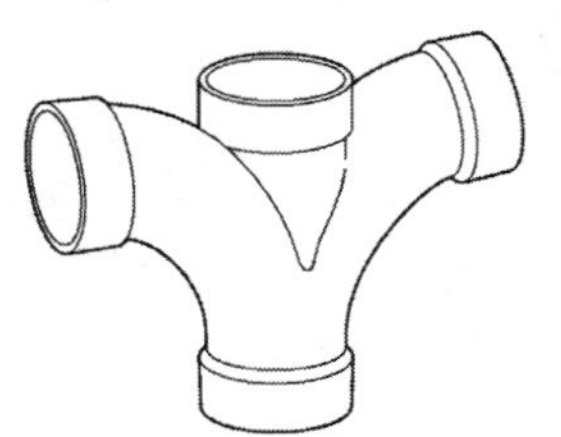

Double Fixture Tee

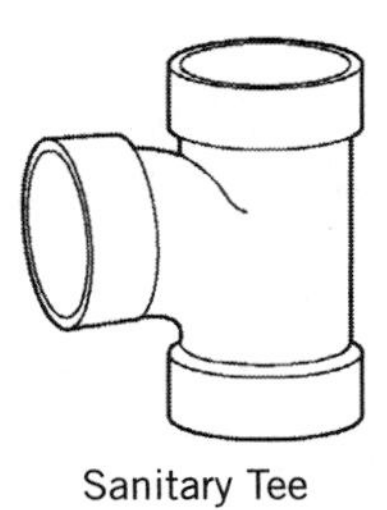

Sanitary Tee

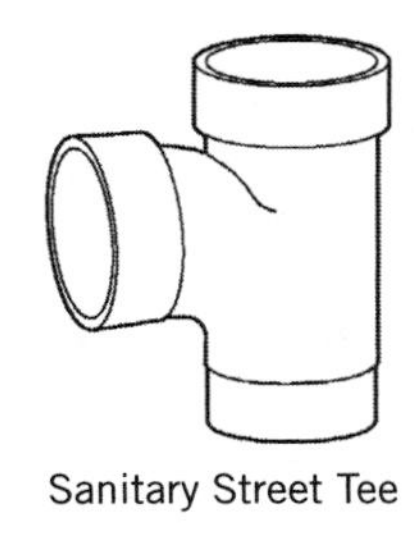

Sanitary Street Tee

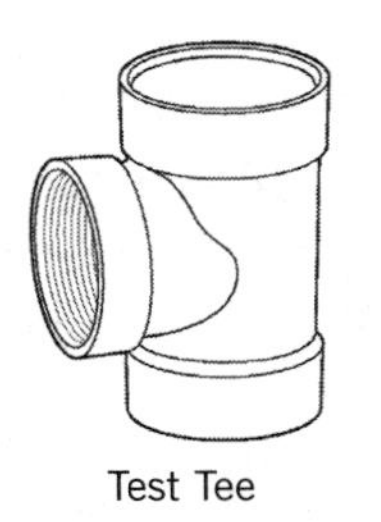

Test Tee

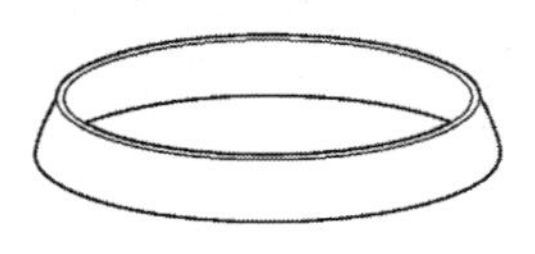

Slip Joint Washer

Figure 13–1 There is a PVC fitting for every need. (concluded)

Courtesy Mueller Industries

The best reason of all is that you can do it yourself.

Figure 13–1 on pages 76–77 shows some of the wide variety of PVC and ABS DWV fittings and their names to help you to know what to ask for.

Installing PVC Pipe and Fittings

Remodeling projects, home improvement, and common plumbing repairs often require the installation of new drainage and water piping. PVC-DWV fittings are used for drainage, waste, and vent piping, CPVC is coming into general use as water piping, and schedule 40 PVC is being used in many areas for water service piping.

The first step in installing piping is to make accurate measurements. Chapter 4, "Measuring and Cutting Pipe," explains end-to-end and end-to-center measurements and makeup points of fittings, and so on.

A hacksaw, Figure 13–2, a ratchet-type tubing cutter, Figure 13–3, or a fine-toothed hand saw, Figure 13–4, can be used to cut the pipe to length. If a saw is used, a miter box should also be used to ensure square cuts.

When cuts are made, burrs on both the inside and outside of the pipe should be removed. The tools shown in Figure 13–5 will remove inside and outside burrs. The outside of the pipe and the socket of the fitting should be wiped clean with a rag.

To ensure proper penetration and fusion of solvent cement when joining PVC components, a *primer* must be used. The primer should be applied first to the inside socket surface, using a scrubbing motion. Repeated applications may be needed to soften this surface. Then apply a liberal portion of the

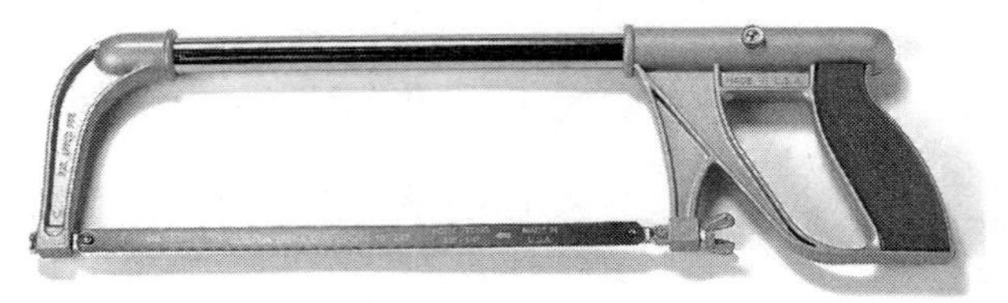

Figure 13–2 A hacksaw.
Courtesy Ridge Tool Co.

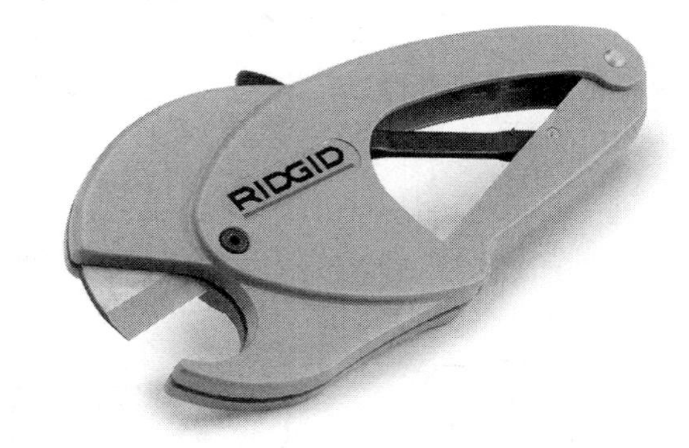

Figure 13–3 A ratchet-type plastic pipe cutter.
Courtesy Ridge Tool Co.

Figure 13–4 A fine-toothed hand saw.
Courtesy Ridge Tool Co.

Figure 13–5 These tools will remove inside and outside burrs on plastic pipe.
Courtesy Ridge Tool Co.

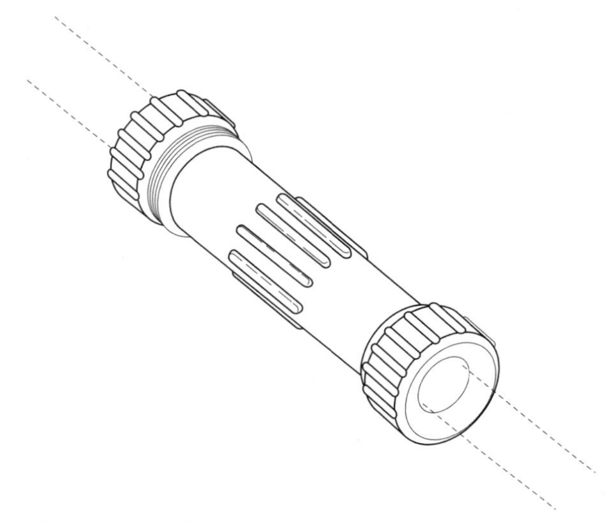

Figure 13–6 Cementing joints in plastic piping.

primer to that portion of the male end of the pipe that will enter the fitting socket. While the socket is still wet, apply the cement to the socket and to the cleaned area of the pipe and quickly insert the pipe into the socket. Turn the pipe a quarter turn while inserting it into the socket, and when it has bottomed out, hold it in place for 30 seconds to prevent the pipe from backing out. When joining pipe and fittings, if the fitting must be turned to an exact direction or angle, this must be done *immediately,* within 3 or 4 seconds.

Applying the primer and cement is shown in Figure 13–6. After the cement has taken its initial set, the joint cannot be turned. Both the primer and the cement have a dauber built into the can's cap. If the dauber is missing, a small soft bristle brush can be used to apply the primer and the cement. Quite often, it is necessary to join PVC-DWV piping to ABS fittings. PVC *primer* should be used for PVC pipe, and ABS *cleaner* for ABS fittings. Multipurpose cement is used to join PVC-DWV and ABS-DWV piping and fittings.

Repairing PVC Pipe

One of the problems when PVC pipe is used is the repair of cracked or broken sections of pipe, particularly if the pipe is in a trench. This used to require digging back long sections of pipe. A Dresser coupling, Figure 13–7, makes this repair easy. A coupling nut on each end of the coupling is unscrewed, the nut is slid over the pipe, a gasket on each end of the coupling is then slid over the pipe, and the body of the coupling is inserted over the broken pipe. When the coupling nuts are tightened, the rubber gaskets are forced into place, and the repair is

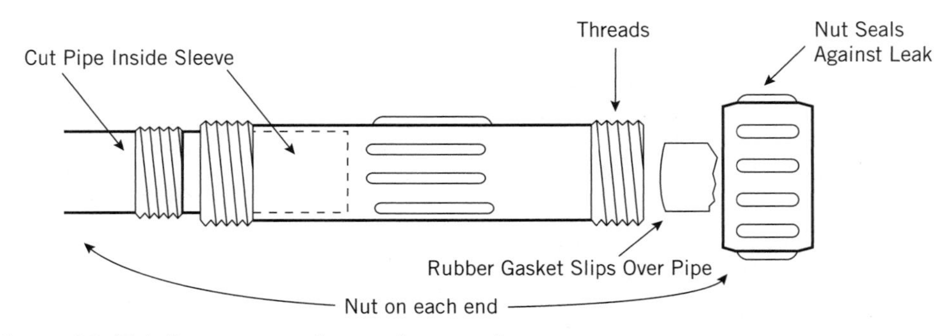

Figure 13–7 A Dresser coupling makes repairs easy.

completed. Also, a telescoping PVC glue coupling is now available. This really comes in handy on a repair job when neither the pipe nor the fitting can be moved.

CPVC Hot- and Cold-Water Piping

It is easy to identify CPVC pipe and fittings; they are tan in color. Because most of the system design parameters, such as minimum pressure, fixture unit, and flow sizing of pipe and limiting velocity, are prescribed in the applicable plumbing code and in ASTM D2846, CPVC tube is usually used as a direct *size-for-size* replacement for copper tube. Because CPVC is a thermoplastic rather than a metal, there are certain differences in handling, cutting, joining, and installation that are detailed here. CPVC pipe and fittings should either be covered or stored indoors.

The American National Standards Institute (ANSI), the American Society for Testing Materials (ASTM), and the National Science Foundation (NSF) rate and test materials. The basis for rating and testing is called standards. Materials that comply with their standards are identified as follows:

- CPVC *pipe* shall have the following information printed on it: the manufacturer's name, certification or listing agency mark, ASTM standard number F441, size, Sch. 80, and pressure rating.
- CPVC *fittings* shall have molded markings of manufacturer's name, certification or listing agency mark, ASTM standard number (D2846 or F439), and material designation (CPVC4120 or CPVC 23447).
- CPVC *solvent cement* shall have on the label CPVC Solvent Cement, ASTM F493, certification listing agency mark, and manufacturer's name.
- *Primer or cleaner* shall have on the label primer or cleaner, ASTM standard number, certification or listing agency mark, and manufacturer's name.

A product that does not have legible marking or has a marking that does not contain all pertinent information may not conform with the applicable standard.

Verify Local Code Approval

CPVC piping is included in all major model plumbing codes: BOCA National Plumbing Code, SBCCI Standard Plumbing Code, IAPMO Uniform Plumbing Code, ICC International Plumbing Code, The BOCA One & Two Family Dwelling Code, The National Standard Plumbing Code (NAPHCC), and in FHA/HUD Bulletins.

NOTE: State and local governments can adopt these model codes as published or modify them. Ask your local building official: Is a model code being used? If so, which one and what if any modifications are in effect in regard to CPVC piping?

NOTE: *Shall* is a mandatory term when used in plumbing codes.

Cutting and Beveling

CPVC can be cut with the tools shown in Figure 13–2, Figure 13–3, and Figure 13–4. Ratchet cutter blades should be sharpened frequently. If a roller-type cutter is used, it should be made for cutting CPVC. Chamfer the cut ends and remove inside burrs. Use a clean dry rag to remove any moisture from the pipe ends and the fitting socket.

Primer and Cement

CPVC pipe and fittings are joined with CPVC cements. The solvent cement can be either a one- or a two-step process. The one-step cement does not require the use of a primer or a cleaner. This process uses a yellow-colored cement. The two-step process requires the use of a primer or cleaner; the cement is orange in color. Both types of cement are manufactured under the ASTM F493 standard for use with CPVC hot- and cold-water piping. The label on the can will indicate the color of its contents and whether or not a primer is required.

If primer is required, apply it to the outer surface of the pipe end and the inner surface of the fitting socket. Apply a light coat of CPVC cement to the socket surface and a heavy coat to the pipe end. Immediately insert the pipe end into the socket, giving it a quarter turn as it bottoms out. Hold the pipe firmly in the socket for at least 10 or 15 seconds to prevent the pipe from pushing out of the socket. If a push out occurs, increase the holding time. If the surface dries before the joint is put together, apply another light coat of cement to the pipe end and then assemble.

Do not use excessive amounts of cement or primer or allow them to puddle in the socket. A good job of cementing is evidenced by an even bead or fillet of cement all around the pipe at the socket interface. Wipe off any excess cement.

Solvent set and cure times are a function of pipe size, temperature, and relative humidity. Follow the solvent cement manufacturer's recommended drying times.

Safe Handling of Primer, Cleaner, and Cement

Avoid prolonged breathing of solvent vapors. When pipe and fittings are being joined in a partially or completely enclosed area, use a ventilating device, fans, and so on, in order to maintain a safe level of vapor concentration with respect to toxicity and flammability in the breathing area. Ventilating devices must be located so as to not provide a source of ignition to flammable vapor mixtures.

Keep containers of cements, primers, and cleaners tightly closed except when the product is being used. Dispose of rags and other materials used on the job in an outdoor safety waste receptacle.

Measuring PVC and CPVC Pipe

In Figure 13–8 a length of pipe extends to the center of a 45° elbow. When working with PVC or CPVC, it is important to remember that pipe will not completely enter or bottom out in the dry socket of a valve or a fitting. When the valve or fitting have been coated with cleaner, primer, or PVC or CPVC cement, the pipe will then

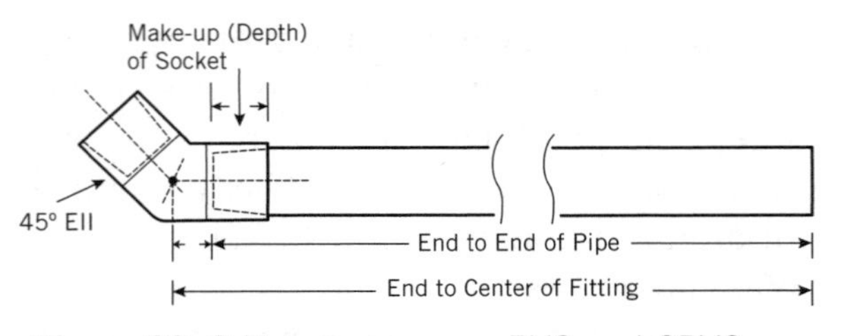

Figure 13–8 How to measure PVC and CPVC pipe and 45° fittings.

enter the full depth of the socket. The most accurate way to measure a length of pipe containing an elbow or a tee is to measure from the end to the center of the fitting.

For example, the length of a pipe needed to reach a point, make a 90° turn, and connect to another pipe is 20 inches. The end of the pipe and the socket of the 90° elbow should be cleaned or primed, and cement should be applied to the fitting socket first, and then to the end of the pipe. The pipe should be inserted into the fitting, turning the pipe one-quarter turn as the fitting is entered. Then, measure from the center of the fitting and mark the pipe at 20 inches. Cut the pipe at this point. When end-to-center measurements are used, it takes the guess-work out. What you see is what you get.

Expansion and Contraction

CPVC pipe expands when heated and contracts when cooled. A 100-foot run of CPVC will expand about 4 inches with a 100°F temperature increase. Expansion does not vary with size. Although expansion joints are available, they are rarely used in water distribution systems in small buildings or residences. Thermal expansion of CPVC systems is usually accommodated at changes in direction or by offsets.

Support Spacing

The following support spacing is recommended for the sizes listed:

Size (inches)	Spacing (inches)
1/2 and 3/4	36 inches
1	40
1 1/4	46
1 1/2	52
2	58

If loads are concentrated, valves and other items provide additional support.

CPVC pipe, tube, and fittings have been successfully used in hot- and cold-water distribution systems since 1960. Millions of single-family homes are now using CPVC piping.

Transition Fittings and Joints

Special transition fittings are used whenever CPVC piping is connected to a metal valve, fitting, or other appurtenances such as a filter or to parts made of another plastic. One common form is the true union with a metal end and a CPVC end held together with a plastic or metal gland nut and having an elastomeric seal between them. Your supplier will have the correct transition fittings to use.

Standard compression fittings that use brass or plastic ferrules can be used to assemble CPVC. If these fittings are used,

Teflon® tape should be applied over the brass ferrule to compensate for the dissimilar thermal expansion rates of the brass and CPVC that could otherwise result in a leak. Female threaded adapters without an elastomeric seal should never be used. Threaded CPVC fittings with tapered pipe threads (male thread adapters) must be used with a suitable thread sealant. Teflon® tape is the most widely accepted and approved sealant.

Water Heater Connections

Some plumbing codes contain detailed requirements for CPVC connections to gas or electric storage-type water heaters. Check the building and plumbing code in effect in your area for requirements for CPVC connections to water heaters.

14

Notes on Stopped-Up Drains

TOOLS NEEDED

Plungers

Slip-joint pliers

Rarely, a cable drain cleaner

I'm sure you have seen the TV commercial showing a trap on a plumbing fixture. The trap is plugged and water is standing in the sink. Then someone pours a liquid drain cleaner into the sink and *presto,* water gushes out of the sink and the drain is open. If you believe this, I've got a bridge in New York City I'll sell for a very reasonable price. Now, let's look at the facts, the causes of stopped-up drains and how drains can be opened safely without danger to persons or damage to property.

Why Drains Stop Up Inside the Home

Grease, hair, and soap buildup over a period of time are the most common cause of drain stoppages inside the home. The age of the home also is a factor: cast-iron soil pipe, lead pipe, steel pipe, and cast-iron fittings were standard drain pipe material inside the home before PVC-DWV pipe and fittings came into use. Cast-iron soil pipe is rough inside, and poorly made joints often offer resistance to the smooth flow of water-carried wastes. When cast-iron soil pipe was used in houses built on concrete slabs, bricks were often used to support the soil pipe. If, as sometimes happened, the brick supports were knocked over as the concrete slab was poured, the drainage pipe then had a sag in it, and stopped-up drains plagued the home owner forever after.

Lead pipe was used for branch waste lines from kitchen sinks and bathroom lavatories and were connected to cast-iron soil and steel pipes. Although lead pipe was smooth inside, the rough joints between these materials was often an obstacle to smooth flow.

DWV copper was a vast improvement over cast-iron soil pipe and steel pipe waste lines, but DWV copper waste lines are subject to damage from acids in human wastes and their use was curtailed.

Drain Stoppages Outside the Home

Until the early 1980s it was common practice to use vitrified tile to carry the soil and waste products from the home to the sewer. As the ground settles, vitrified tile settles and cracks, allowing dirt and roots to penetrate the tile. Roots are one of the main offenders causing blockages in sewers. If your house has a PVC building drain and sewer, you may never have a stopped-up drain.

If all the drains in the home run slowly or are completely stopped up, you may be certain that the sewer from the home to the main sewer in the street is blocked up. Getting this sewer open is not a job for the average home owner. Your local plumber or drain cleaner has special equipment to do this job. This equipment can cut out roots, and the operator can tell from the way the cable feels as it travels through the tile whether the trouble is caused by roots or a broken tile. If roots are the problem, a copper sulphate solution should be poured into the sewer periodically, preferably through a cleanout opening to slow down or prevent the regrowth of roots. Copper sulphate for this purpose is sold by hardware stores.

If the cable reaches a point and starts to kink and twist, this indicates that the tile is broken and must be replaced. Measuring the length of cable that was inserted in the drain tile can give a pretty fair estimate as to how far from the building the tile is broken. Knowing where the drain line leaves the building and the direction of flow will help to determine approximately the location of the broken tile. Rather than just replacing a broken tile, it would be better to replace the entire run of pipe with PVC pipe.

Kitchen Sink Drains

Kitchen sink drains often stop up because a garbage disposer is worn out and no longer chops vegetables finely enough to go through the sink trap and into the waste pipe. The obvious cure is to install a new disposer. Instructions for installing a garbage disposer are in Chapter 15.

The old fashioned plunger, Figure 14–1, is the most effective tool to use on a stopped-up sink drain. When a plunger is pushed down repeatedly over a drain opening, a shock wave is transmitted through the water in the pipe; continuing use of the plunger, causing a series of shock waves, will often break a stoppage loose. If you have a two-compartment sink, use one plunger to seal the opening on one side while you use the other plunger on the other side. If the plunger does not open the drain, it might just be the time to call a plumber or drain cleaning company. Let someone else do this dirty job.

Figure 14–1 Plungers can open stopped–up drains.

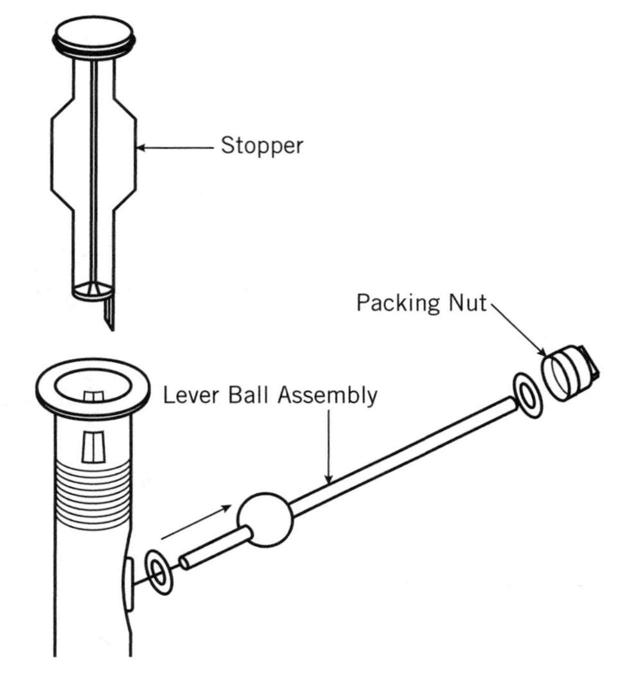

Figure 14–2 Use a plunger after removing the stopper.

Lavatory Drains

The procedure for unstopping a lavatory drain is basically the same as for a kitchen sink drain. The most common cause of a stopped-up lavatory drain is an accumulation of hair. If the lavatory has a pop-up drain, lift out the stopper in the bowl. Some stoppers are locked into the lever ball assembly rod in the pop-up waste. Turn the packing nut securing the lever ball assembly counterclockwise to loosen and remove it, and then pull the lever ball assembly out of the pop-up body. The stopper can now be lifted out of the bowl. Remove the hair that has accumulated at the bottom of the stopper. Use a plunger, Figure 14–2, or wet rag to cover the overflow hole and, using another plunger on the drain opening, work the plunger up and down vigorously. Plungers really work when it comes to unstopping drains. Replace the stopper and reconnect the lever ball assembly.

If using the plunger does not open the drain, place a bucket or large pan under the J-bend of the trap and use your slip-joint pliers to turn the slip nuts on the J-bend counterclockwise to loosen them and remove the J-bend. If the trap arm is connected to the waste piping by a slip nut, turn the nut counterclockwise and pull the trap arm out.

If the trap is connected to a galvanized steel waste pipe by a solder bushing, Figure 14–3, remove the J-bend and insert a small (1/4 or 5/16 inch) cable into the J-bend, Figure 14–4. Spinning the top as the cable is being forced back into the drain should open the drain.

If the trap is connected to a galvanized steel waste pipe by a solder bushing, refer to Chapter 8 for instructions on how to remove and replace the solder bushing.

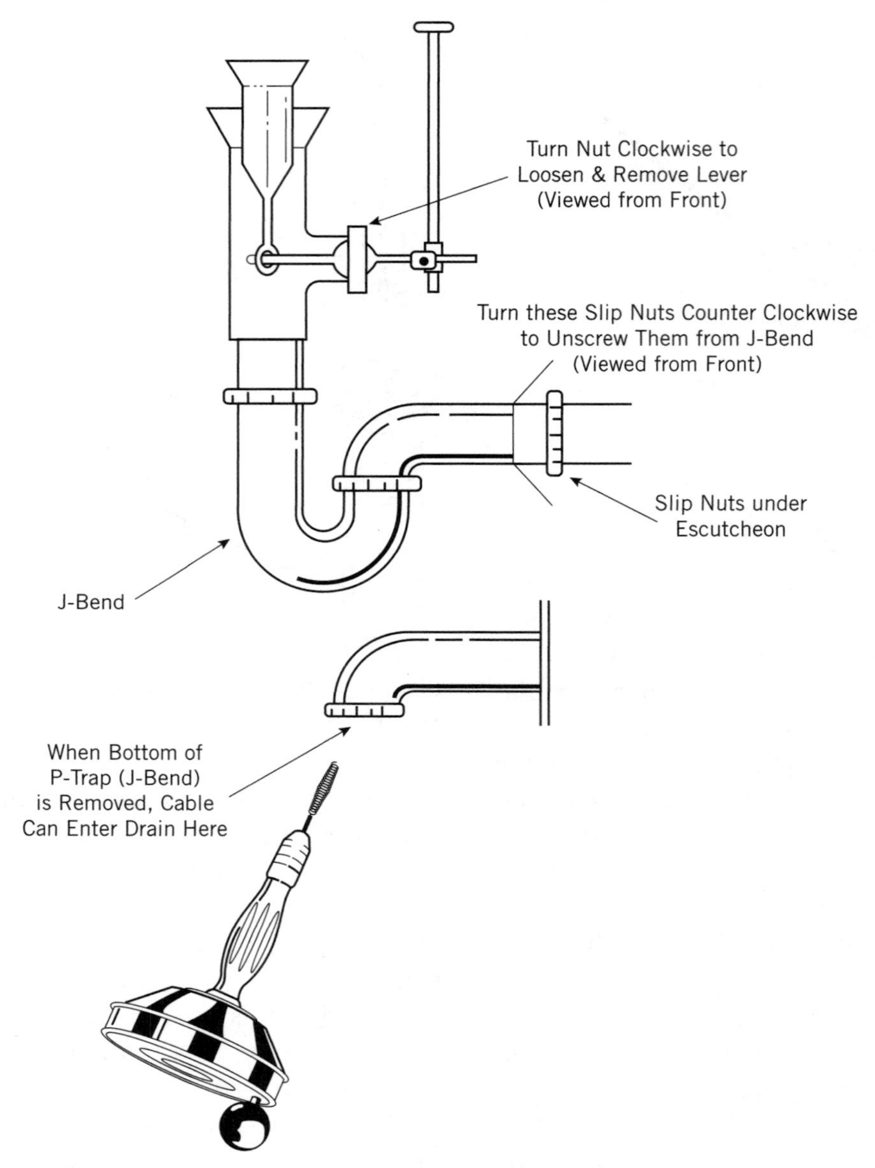

Figure 14–3 Spinning the top helps force cable through a drain.
Courtesy Ridge Tool co.

Toilets and Bidets

Every plumber carries a closet (toilet) auger for that very rare occasion when a customer has a stopped-up toilet. I do not believe the average homeowner will ever need one. The good old plunger, the "plumber's helper," will step in and open a toilet or bidet in an emergency. If you should need a closet auger, you will find it at the local hardware store.

A final word about drain cleaners. Your local grocery and hardware stores have liquid drain cleaners in bottles that promise to open stopped-up drains. They also have chemicals in cans that promise to open stopped-up drains. Some of these contain lye, sodium hydroxide, which when it comes in contact with water, generates heat, a lot of heat. Fixture traps on kitchen sinks and lavatories are now commonly made of PVC, a thermosetting material. These traps are shaped while the material is hot, and when it cools, the traps hold their shape. If lye is poured into the water in a stopped-up kitchen sink or lavatory, the water in it can get so hot that the trap serving the fixture loses its shape and can fall off of the fixture, spilling the hot lye water in the cabinet under the fixture. Lye can cause severe burns on human skin and damage to cabinets, linoleum, and so on. *Beware of these products.*

Figure 14–4 Remove J-bend to allow cable to enter drain.

15

Replacing a Garbage Disposer

TOOLS NEEDED

Large slip-joint pliers

Screwdrivers

10-inch pipe wrench

Small adjustable wrench

Plumbers putty

Small bucket

Today every new home comes equipped with a garbage disposer, but in 3 years or less many of these disposers seem to self-destruct and must be replaced. Builders are not concerned with the life expectancy of the disposers they install; most builders go strictly by price, not quality. They select the least-expensive models, called builders' models in the trade. Home buyers, seeing that the kitchen sink contains a disposer, accept it and assume that the disposer will give years of service.

I have chosen two very well-known name brands of disposers to show you that there are choices when it comes to selecting a new disposer. You will want a quiet yet powerful disposer, one that is not intimidated by a marshmallow. Cold water should always be running when a garbage disposer is in use. Running hot water is not recommended when garbage is being ground. Both brands shown in Figure 15–1, Figure 15–2, Figure 15–3, and Figure 15–4 offer in-home warranties.

The 3/4 HP Model 17® In-Sink-Erator, Figure 15–5, does not require a wall switch. It is what is called a batch feed disposer. When food scraps are put into the grinding chamber, the disposer will operate *only* when the drain cover is inserted. I consider this a safety feature. The Model 17 also has a 5-year in-home service warranty.

Figure 15–1 In–Sink–Erator Badger® 1 disposer.
Courtesy In–Sink–Erator Division/Division of Emerson Electric

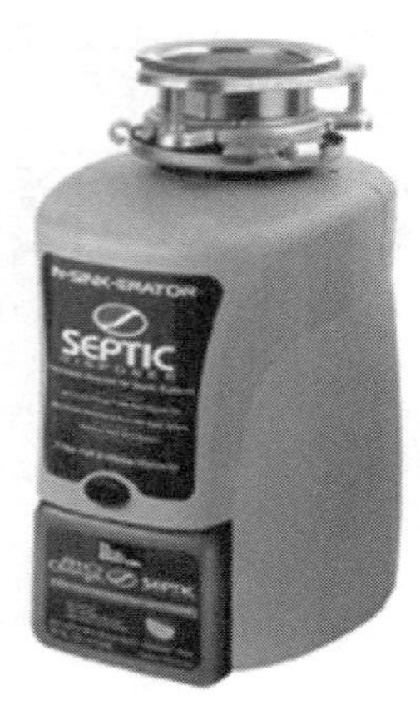

Figure 15–2 In–Sink–Erator Septic disposer.
Courtesy In–Sink–Erator Division/Division of Emerson Electric

Figure 15–3 In–Sink–Erator Model 555ss® 3/4 HP disposer.
Courtesy In–Sink–Erator Division/Division of Emerson Electric

Figure 15–4 In–Sink–Erator Model 777ss® 1 HP disposer.
Courtesy In–Sink–Erator Division/Division of Emerson Electric

All disposers are subject to jams caused by bone fragments or metal objects that lodge in the grinding chamber of a disposer. In-Sink-Erator furnishes a wrenchette, a tool with a 1/4-inch hex-shaped end with each disposer. If the disposer becomes jammed, make certain that the disposer is turned off, and then insert the wrenchette in the opening in the *bottom center* of the disposer. The wrenchette will turn the plate at the bottom of the grinding chamber, freeing the jam.

Also, a thermal overload switch is found on the bottom of disposers. If a disposer does not operate when turned on, it may have been overloaded. Press the red button on the thermal overload to restart the disposer.

The installation instructions furnished with In-Sink-Erators are very clear, and the installation should be a piece of cake.

Figure 15–5 In–Sink–Erator Model17ss® 3/4 HP disposer.
Courtesy In–Sink–Erator Division/Division of Emerson Electric

Figure 15–6 The Waste King Model 111 disposer.
Courtesy Waste King

Figure 15–7 The Waste King Model 2600 disposer.
Courtesy Waste King

Figure 15–8 The Waste King Model 3100 disposer.
Courtesy Waste King

Figure 15–9 The Waste King Model 8000 disposer.
Courtesy Waste King

Figure 15–10 The Waste King Model 8000TC disposer.
Courtesy Waste King

CAUTION: If you are connecting a dishwasher drain to the disposer, remember to remove the plug in the dishwasher drain hole opening. Check your installation instructions.

Waste King

Waste King offers eight models to choose from ranging from 1/3 to 1 horsepower (hp). I have included five of the Waste King models. The step-by-step installation instructions furnished by Waste King are excellent. All models carry an in-home service warranty. The warranty varies from the 2-year warranty on the 1/3-horsepower Model 111, Figure 15–6, to the lifetime warranty on the Model 8000.

The 1/2-horsepower Model 2600, Figure 15–7, and 1/2-horsepower Model 3100, Figure 15–8, are the middle of the line Waste King disposers. The 2600 carries a 5-year warranty, and the 3100 carries an 8-year warranty.

The 1-horsepower Waste King Model 8000, Figure 15–9, carries a lifetime warranty.

The 1-horsepower Waste King Model 8000TC, Figure 15–10, is a batch-feed disposer, as is the 5000TC, and carries a 10-year warranty.

All models feature the fast and easy mounting system, and many models have the power cord installed. Waste King features 180° swivel impellers to virtually eliminate jamming.

16

Replacing a Residential Dishwasher

TOOLS NEEDED

Slip-joint pliers

6- and 8-inch adjustable wrenches

Screwdrivers

Scraps of carpet or cardboard

MATERIALS NEEDED

Dishwasher

Braided flexible dishwasher connector

Air gap fitting

Wirenuts

Repair or Replace?

Should you repair or replace that old dishwasher? Every few years this question comes up in every home. There was a time when I might have said repair it, but now I say buy a new one. We live in a throwaway world. The cost of a motor or a pump plus the labor involved would probably equal or exceed the cost of a new unit. Replacing a dishwasher is usually much easier than a new installation because, unless the dishwasher is very old, it is simply a matter of removing the old unit, making some minor changes, and connecting the new one.

When you are looking at new dishwashers, also look for a braided flexible stainless steel dishwasher connection. If your old dishwasher uses soft copper tubing for the water supply to the dishwasher, it will be hard to work with when connecting it to the new unit. Do it the easy way; use a braided flexible stainless steel connector. This connector is available in several different lengths. If you are in doubt as to which length to get, wait until you have removed the old copper water supply line and then get the new flexible supply.

You will receive installation instructions with your new unit, telling you where the water and drain connections are and where to connect the electric wiring. In this chapter I will tell you some things that will make the installation much easier plus some details the installation instructions don't cover.

If you are replacing an existing dishwasher, the water and electrical connections should already be in place. Changes in electrical codes may require that the present connection points and controls be brought up to today's standards. It is a good idea to have a licensed electrician check this out. Another point to check is the drain hose connection to a sink tailpiece or garbage disposer. The installation instructions for many dishwashers show a looped drain hose in lieu of an air gap fitting. The looped drain hose does not comply with plumbing codes.

The plumbing codes and regulations in effect in the area where work is being done must be observed.

Removing the Old Unit

The first step in replacing an existing dishwasher is to turn off the electricity to the unit. Turn off the circuit breaker or remove the fuse controlling the circuit to the dishwasher in the main electrical panel. Then turn the dishwasher *on* to make certain that the electricity is off. Place a strip of tape over the circuit breaker to prevent someone from turning the breaker to the ON position before the installation is completed.

It is common practice for the control valves for the dishwasher and the kitchen sink faucet to be located under the countertop as shown in Figure 16–1. They may be arranged differently, but they serve the same purpose. The control valve for the hot-water connections to the sink faucet and the dishwasher should be on the left. These valves may be a little hard to get to, behind the garbage disposer and the sink waste piping. Turn the handle of the control valve supplying hot water to the old dishwasher clockwise to turn the water off. Turn the coupling nut connecting the copper tubing to the control valve counterclockwise. If the valve is shut off completely, a few drops of water will drip out. If water continues to flow, turn the valve off again. If the control valve cannot be shut off, refer to Chapter 3 in this book.

Removing the bottom panel on the front of the unit will permit access to the water and electrical connections to the unit. The copper tubing connection to the solenoid valve is made with a coupling nut. Turn this nut counterclockwise to remove it. Leave the copper tubing where it is for now.

Have a pan or bucket ready to catch water and disconnect the drain hose from the garbage disposer or sink tailpiece. The old hose will come out as the dishwasher is pulled out.

Open the dishwasher door. At the top of the opening two screws should secure the unit to the countertop. Remove these screws. The unit is held against the countertop by four leveling legs on the bottom of the unit. The legs, one at each corner, have adjusting screws. Turn these screws counterclockwise to run them in, lowering the unit away from the countertop. Place scraps of carpet or cardboard under the leveling legs to protect the kitchen floor before sliding the old dishwasher out from under the counter. Disconnect the three wires at the solenoid connection, a white wire (neutral), a black or colored wire (hot), and a green or bare

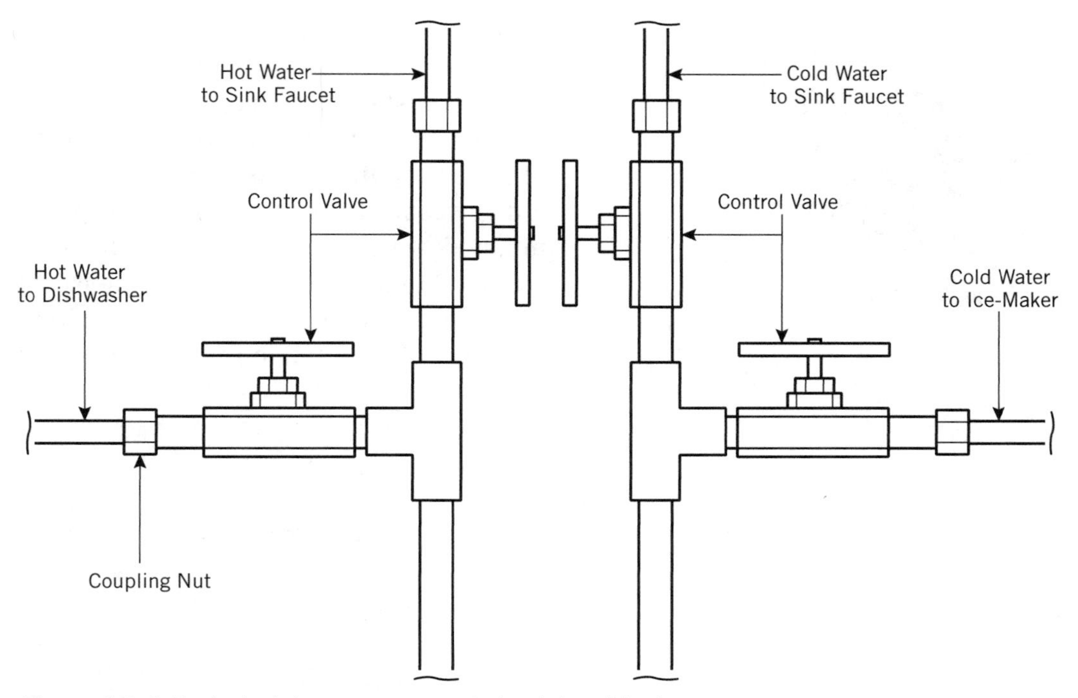

Figure 16–1 Typical piping arrangements in sink cabinet.

wire (ground), and save the wirenuts. The green or bare wire should be connected by a green screw to the metal box housing the solenoid valve. Now we are ready to slide the old dishwasher out and dispose of it. With the old dishwasher out, we are ready to install the new one. There are preparations to be made that will make the installation much easier.

Installing the New Unit

Unpack the new unit and place it on cardboard or a throw rug to avoid damage to the floor. The installation instructions furnished with the unit will specify where the plumbing and electric services must enter the dishwasher area. Because you are replacing an existing unit, these connections will already be there.

Disconnect the copper water line from the valve that controls water to the dishwasher. Although the old copper tubing can be connected to the new dishwasher, do yourself a big favor. Throw the copper tubing away and buy a new flexible dishwasher connector, Figure 16–2. They are available in lengths to fit your needs. Connect one end of the new connector to the control valve and leave the other end where you can reach it when the new unit is in place.

Before sliding the new dishwasher into place, connect one end of the drain hose to the dishwasher pump. Insert the other end

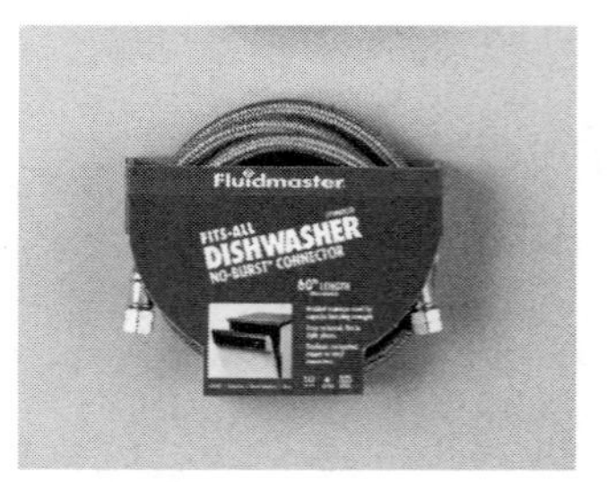

Figure 16–2 Flexible stainless steel dishwasher connection.
Courtesy Fluidmaster

of the hose into the undersink cabinet area. Take the slack out of the hose as you slide the dishwasher into place.

Plumbing codes require the installation of an air gap in the waste water piping from a dishwasher to prevent backflow of contaminated water from the sink into a dishwasher. Some manufacturers in their installation instructions show a high drain loop as an optional alternative to an air gap in the

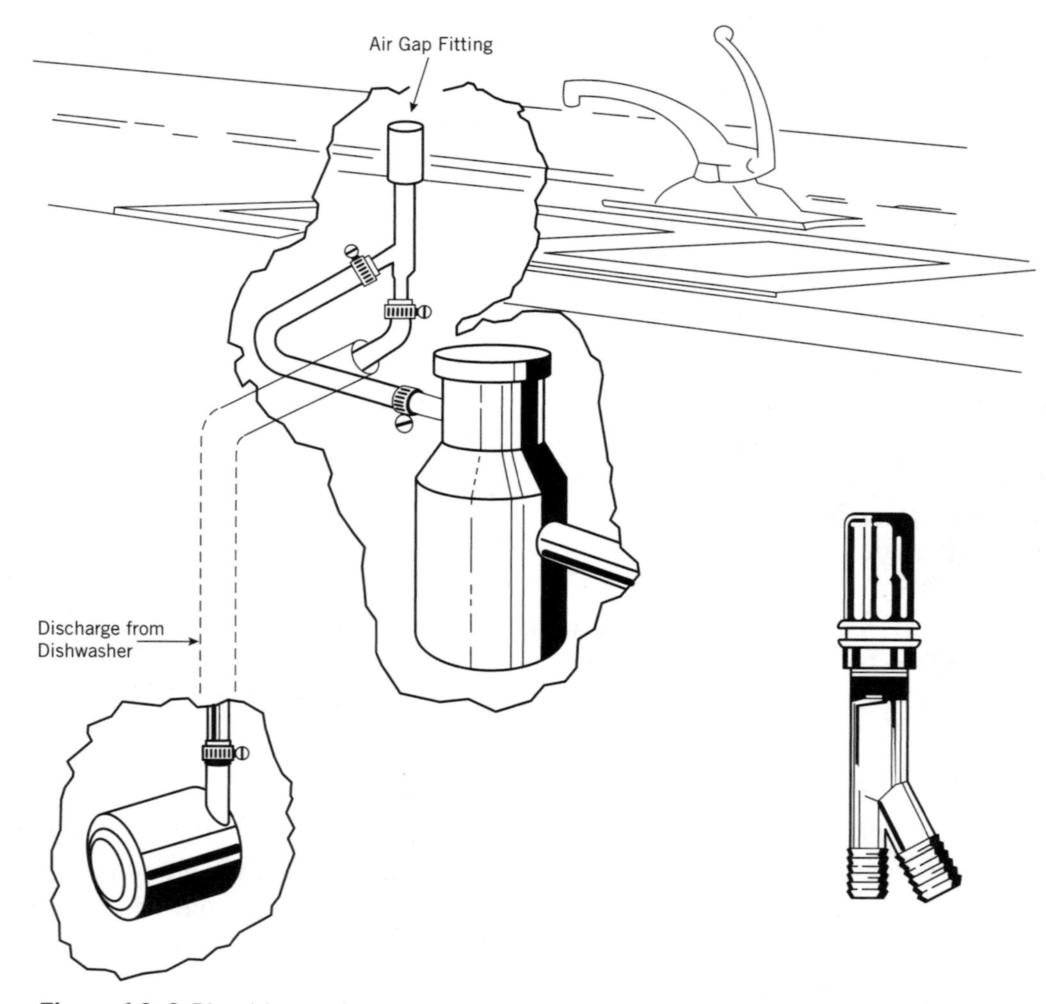

Figure 16–3 Plumbing codes require an air gap fitting in dishwasher drain piping.

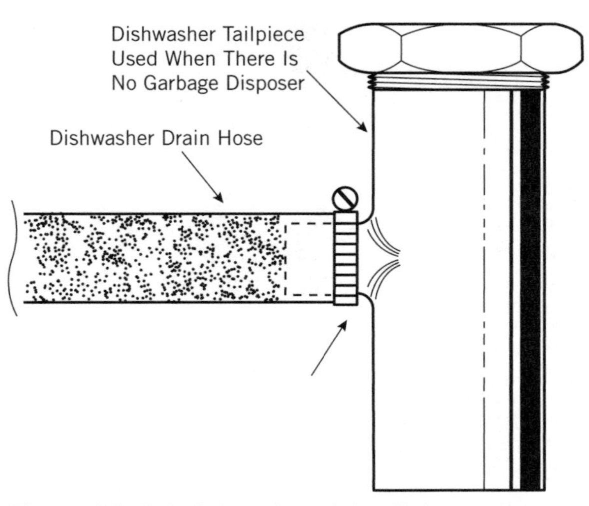

Figure 16–4 A dishwasher sink tailpiece.

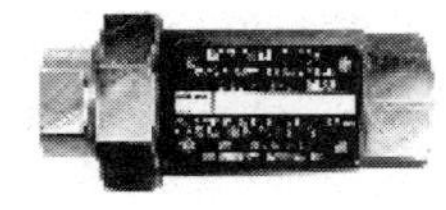

Figure 16–5 A 3/8-inch Watts 7C backflow preventer.
Courtesy Watts Regulator

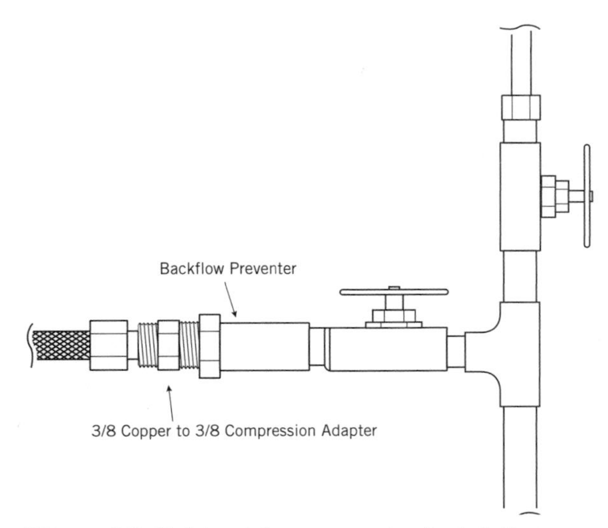

Figure 16–6 A backflow preventer installation.

drain line from the dishwasher to the garbage disposer or sink tailpiece. The looped high drain hose does not comply with plumbing codes.

Figure 16–3 shows an air gap fitting. The drain hose is routed to the inlet of the air gap and from the outlet of the air gap to the disposer or dishwasher tailpiece as shown in Figure 16–4.

The air gap can be installed near the rear edge of the sink rim, in the sink cabinet. The hose connections are in the sink cabinet, and only the chrome top of the air gap is visible above the countertop.

Local plumbing codes may also require the installation of a backflow preventer in the hot-water supply to the dishwasher. This backflow preventer is designed to prevent backflow of contaminated water from the dishwasher into the domestic hot-water supply.

Building officials and plumbing inspectors in various areas of the country may differ as to the exact type or kind of backflow preventer to install; I recommend the Watts 7C, Figure 16–5, as the most satisfactory device for backflow prevention in the hot-water supply to the dishwasher. Check this out with your local building official.

The water connection to the solenoid valve on most, if not all, dishwashers has a 3/8-inch F.I.P. thread. A straight 3/8-inch M.I.P to 3/8-inch compression adapter is used here, and the braided flexible stainless steel dishwasher connector will fit the 3/8-inch adapter. The compression nut on a braided stainless steel dishwasher connector will fit the adapter fitting shown in Figure 16–6.

When installation is completed following the manufacturer's instructions, the water supply can be turned on and tested and the electric power turned on. Now all that remains to be done is to put the bottom panel on the dishwasher and the job is finished.

NOTE: All plumbing and electrical codes and regulations must be followed when installing a dishwasher.

17

Working with Cast-Iron Soil Pipe

NOTE: The ever-increasing use of plastics threatens to eventually eliminate one of the oldest skills required in plumbing, installation of cast-iron soil pipe with lead and oakum joints. To all intents and purposes lead work, use of lead pipe with wiped joints, has gone the way of the dinosaur. Before this happens with lead and oakum soil pipe, this chapter explains how to install cast-iron soil pipe with lead and oakum joints.

Facts about Cast-Iron Soil Pipe

Cast-iron soil pipe is made in two grades, standard or service weight, and extra heavy. It is very durable and is widely used for drainage, waste, and vent piping. After casting, both pipe and fittings are dipped in a special coating for rustproofing.

Cast-iron soil pipe is made to be used with lead and oakum joints and is also available in a No-Hub type. Use of No-Hub soil pipe is self-explanatory. Lead and oakum joints require skill in installation and are explained here.

Cast iron has the characteristic of rusting on the surface while at the same time protecting itself against the continued rusting that would destroy the metal in time. The initial coat of rust adheres strongly to the metal and prevents further rusting.

Single hub (S.H.) soil pipe is made in 5- and 10-foot lengths. Double hub (D.H.) is made in 2- and 5-foot lengths. Some common soil pipe fittings and their names are shown in Figure 17–1.

Cast-iron soil pipe is one of the most durable materials available for drainage, waste, and vent piping. Special tools are needed when working with soil pipe. Cold chisels or soil pipe cutters may be used to cut cast-iron pipe to the desired length. Safety eyeglasses should always be worn when working with cast-iron soil pipe. A yarning iron, packing iron, hammer, and inside and outside caulking irons are needed to make the joint between two pieces of soil pipe. A lead-melting furnace is used to melt the lead. When the lead is hot, a ladle is used to pour lead into the joint. When horizontal joints are made, a joint runner (a braided asbestos rope soaked in oil to soften it) is placed around the joint to hold the molten lead in place until the lead has solidified. Molten lead is dangerous to work with, but some simple precautions will minimize the danger.

Never add lead that is wet or even slightly damp to a pot of molten lead, and *never* put a wet or even slightly damp ladle into a pot of molten lead. Hold the ladle above the pot of melted lead until it is dry. When lead has to be added to the pot, lay a cake of lead in the ladle and hold the ladle over the pot until all moisture has evaporated from the ladle and the lead. If a damp or wet ladle or cake of lead is inserted into the molten lead in the pot, some of the moisture will be carried under the surface of the molten lead. The moisture will be instantly converted into steam. The steam then literally explodes, throwing hundreds of minuscule particles of molten lead in all directions. Severe burns and loss of eyesight are only two of the serious consequences that could occur in such an accident. And, once again, always wear safety eyeglasses when working with cast-iron soil pipe or hot lead.

Never pour melted lead into a wet joint or a joint in which the oakum is wet or damp. Water under the hot lead will convert into steam and an explosion could result. Use only oiled or tarred oakum in soil pipe joints. Oiled or tarred oakum resists moisture. For small jobs oakum is generally sold in 5-pound boxes; the strands are approximately 2 feet in length. Several strands are twisted together; the oakum should be untwisted and used one strand at a time.

How to Measure Soil Pipe Cuts

For accuracy, always use center-to-center measurements. Different terms are used for soil pipe fittings. For example, elbows are called bends. Thus a soil pipe fitting making a 90° turn is called a 1/4 bend. A 45° fitting is an 1/8 bend (see Figure 17–2).

In Figure 17–3 a piece of 4-inch soil pipe must be 42 inches from end (of pipe) to center of a 1/4 bend. A standard (short) 4-inch 1/4 bend measures 8 inches from end of spigot to the center of hub. Deducting 8 inches from 42 inches gives us 34 inches as the length of soil pipe needed. The way to mea-

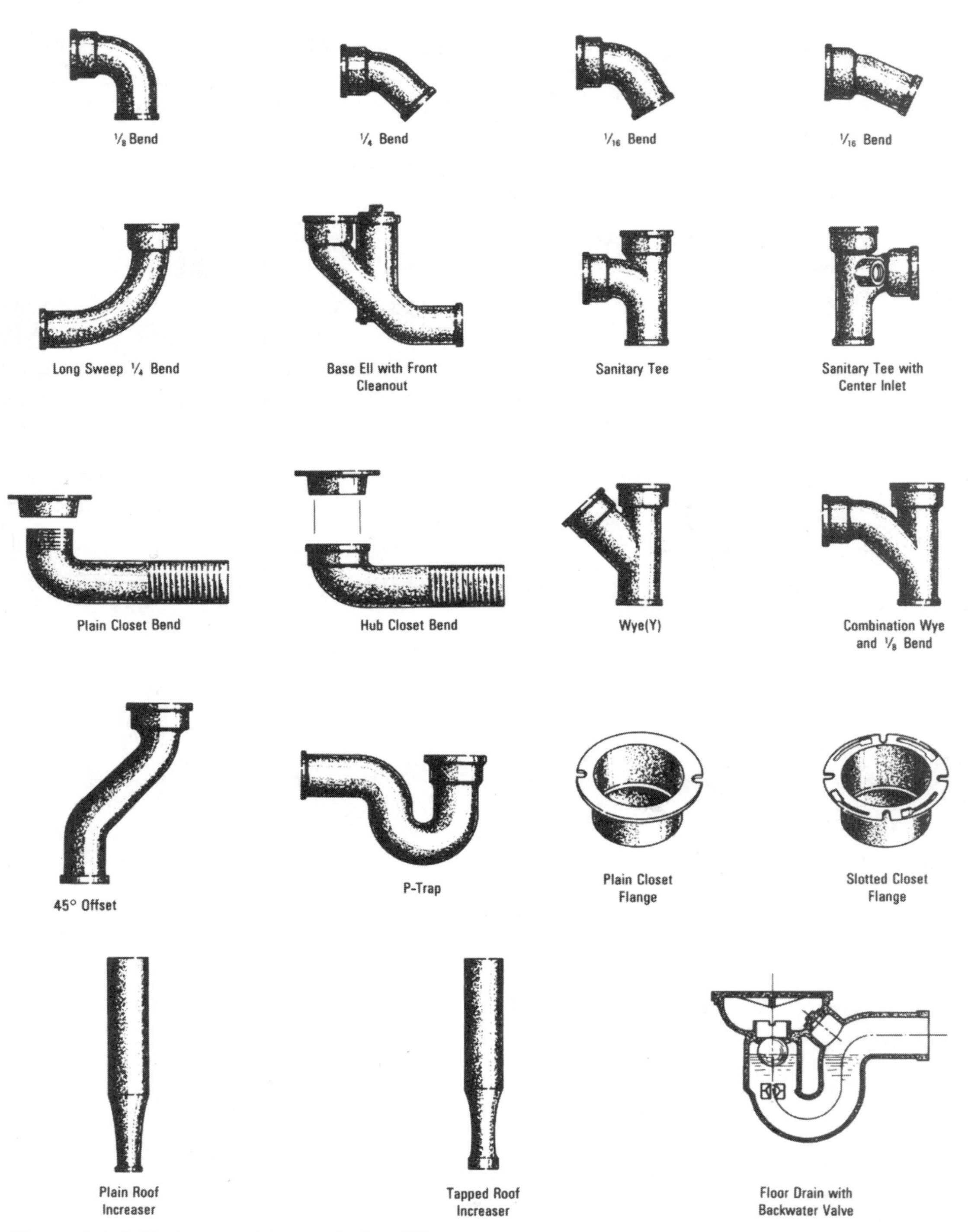

Figure 17–1 Various cast iron soil pipe fittings.

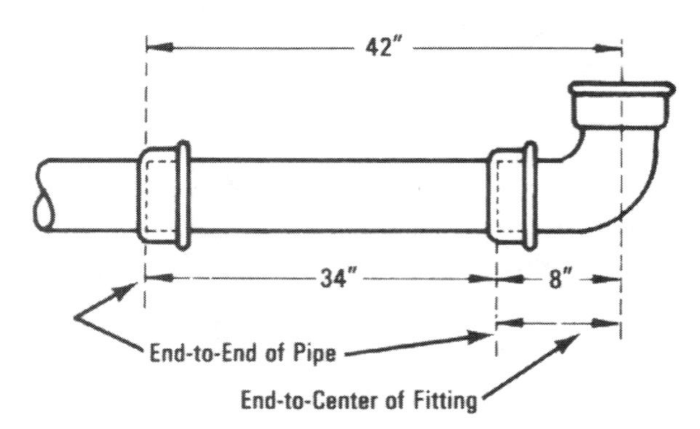

Figure 17–2 How to measure soil pipe fittings.

sure soil pipe fittings is shown in Figure 17–2. The pipe can be cut using soil pipe cutters, or it can be cut as shown in Figure 17–3.

Cast-iron soil pipe can be joined in several ways, but the conventional way is to use lead and oakum joints.

Making a Lead Joint

Steps in making a lead joint are shown in Figure 17–4. When making a joint, insert the spigot end of the pipe or fitting into a hub. Start a strand of oakum at the top of

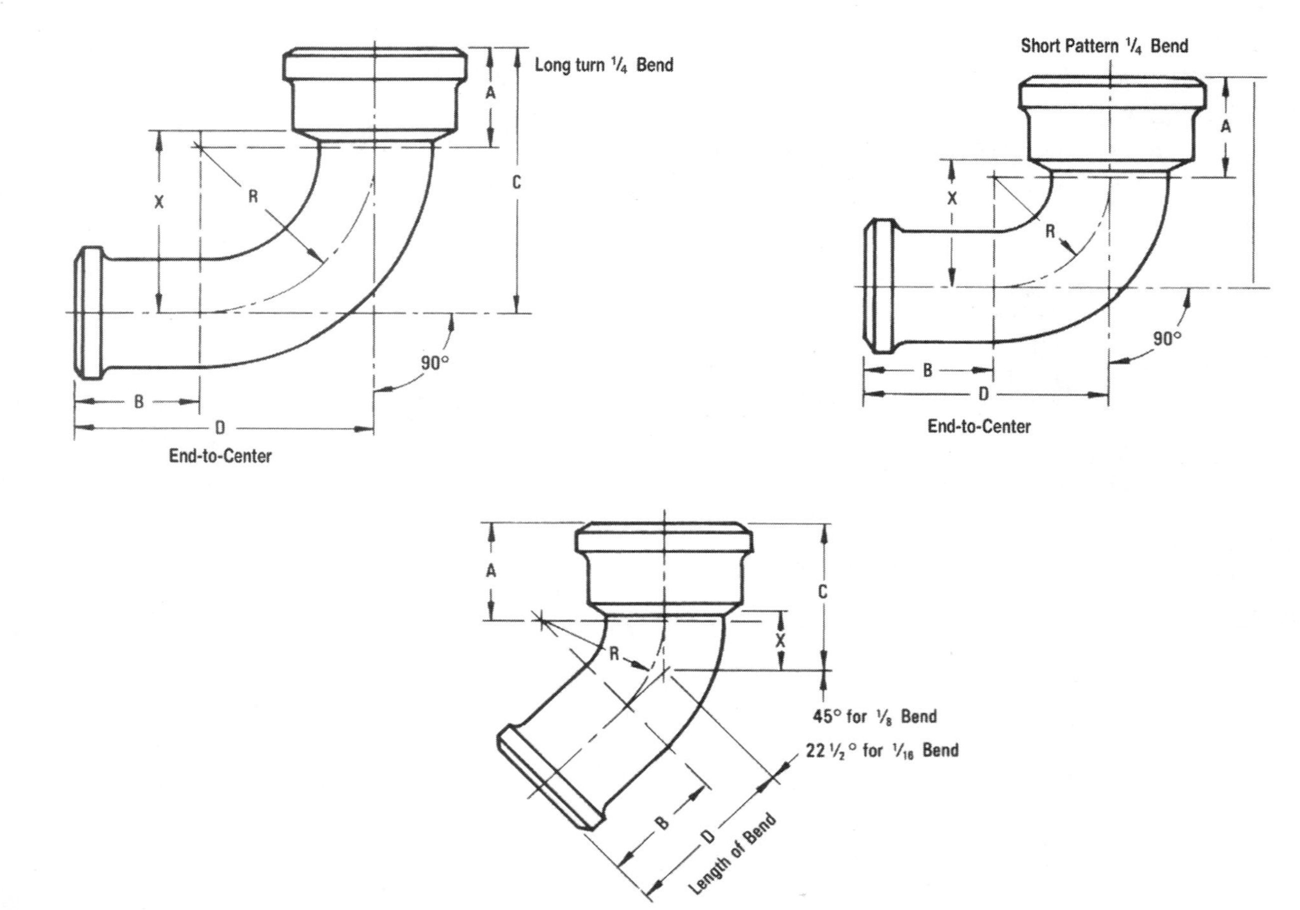

Figure 17–3 How to measure soil pipe cuts.

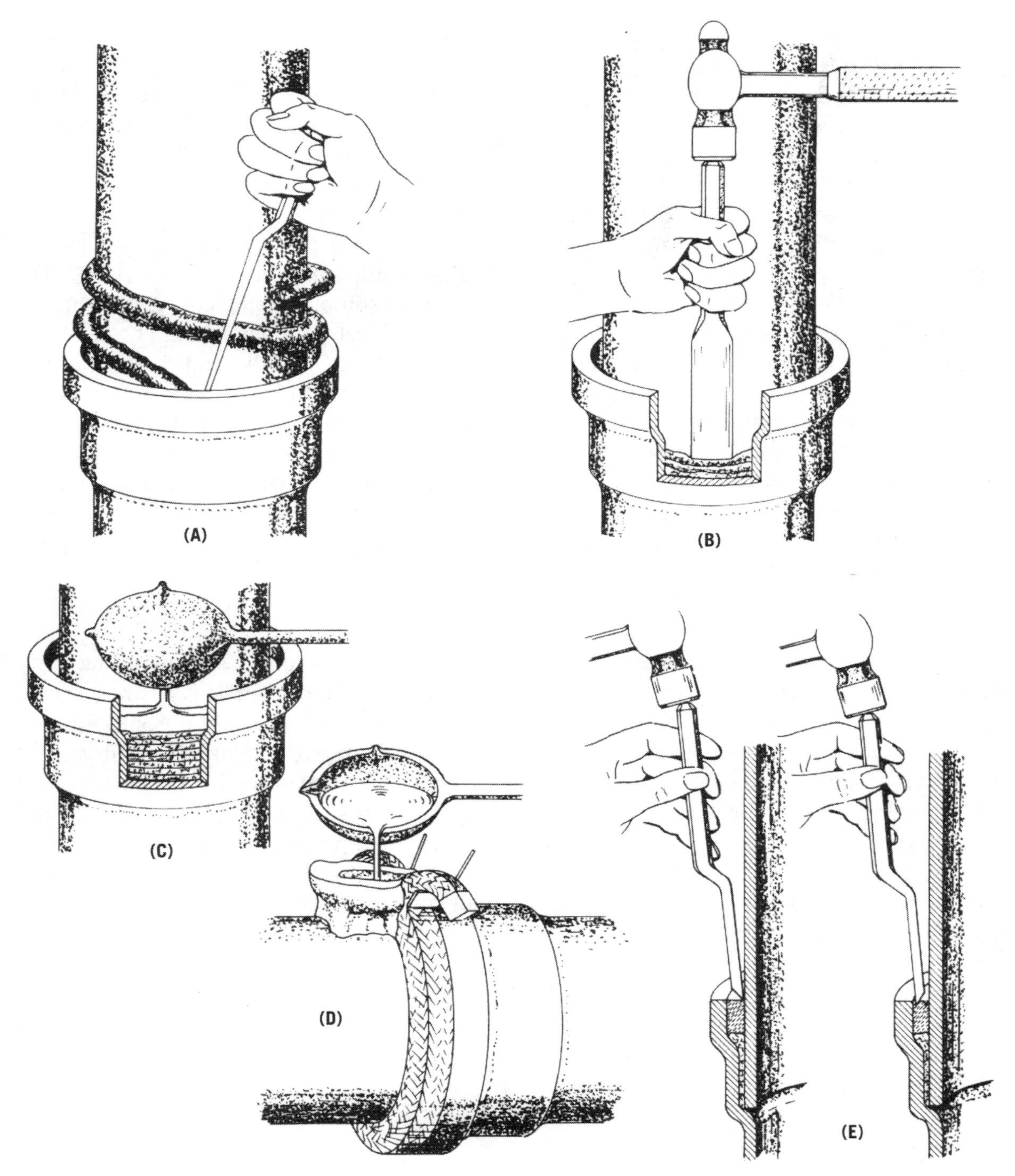

Figure 17–4 Steps in making lead joints.

the hub, (A), pushing the strand down into the hub with the yarning iron. Follow the strand of oakum around the hub, ramming the oakum into place with the yarning iron. Pack the oakum tightly in the hub using a hammer and packing iron, (B). To make a 4-inch joint, the oakum should be approximately 3/4 inch below the top of the hub when the oakum is packed. For 2- to 3-inch joints, figure approximately 1/2 to 3/4 inch. Then pour the lead into the joint, (C). You should fill the joint to the top in one pour. The lead will solidify almost immediately. Use a light hammer to caulk the joint. There are two important points to remember when caulking a joint: Use a lightweight hammer, not over 12 ounces, and do not hit the caulking iron too hard. Lead is soft and when too much force is used on the caulking irons, the lead will be expanded against the hub with enough force to crack it.

The oakum, not the lead, is the key to a good joint. The old joke among plumbers, "Do a good job, use lots of oakum" (lead was much more expensive than oakum), was not really a joke after all. If the oakum is tightly packed in the joint, the joint will not leak. The lead only serves to keep the oakum in place.

To recap, in Figure 17–4 (A) oakum is being rammed into place with the yarning iron. In Figure 17–4 (B) a hammer and packing iron tighten up the oakum. In Figure 17–4 (C) one pour should fill the joint to the top. In Figure 17–4 (D) a joint runner is used to pour a horizontal joint. The gate, made of moist earth or putty, seals the area where the lead is poured. Figure 17–4 (E) shows how the inside and outside irons are used to caulk the lead in the joint.

Figure 17–5 shows tools needed when working with soil pipe. In Figure 17–5 (K) we see the soil pipe being cut with hammer and cold chisel. The pipe should be marked with soapstone or chalk, and the cold chisel should follow this mark to make an even cut. A wooden block should be placed under the cut to provide a solid base for the pipe. When a hammer and chisel is used to cut pipe, safety glasses should be worn for eye protection.

Figure 17–4 (D) shows how to pour a horizontal joint. Prepare the joint just as a vertical joint, inserting and packing the oakum. After soaking the joint runner (an asbestos rope) in oil until it is flexible, place it around the joint and hold it in place with a clamp. Find some damp soil (putty will do) and make a gate with it. Pour the hot lead into the gate. It will flow around the joint and set up (harden) almost immediately. Then release the clamp and remove the joint runner.

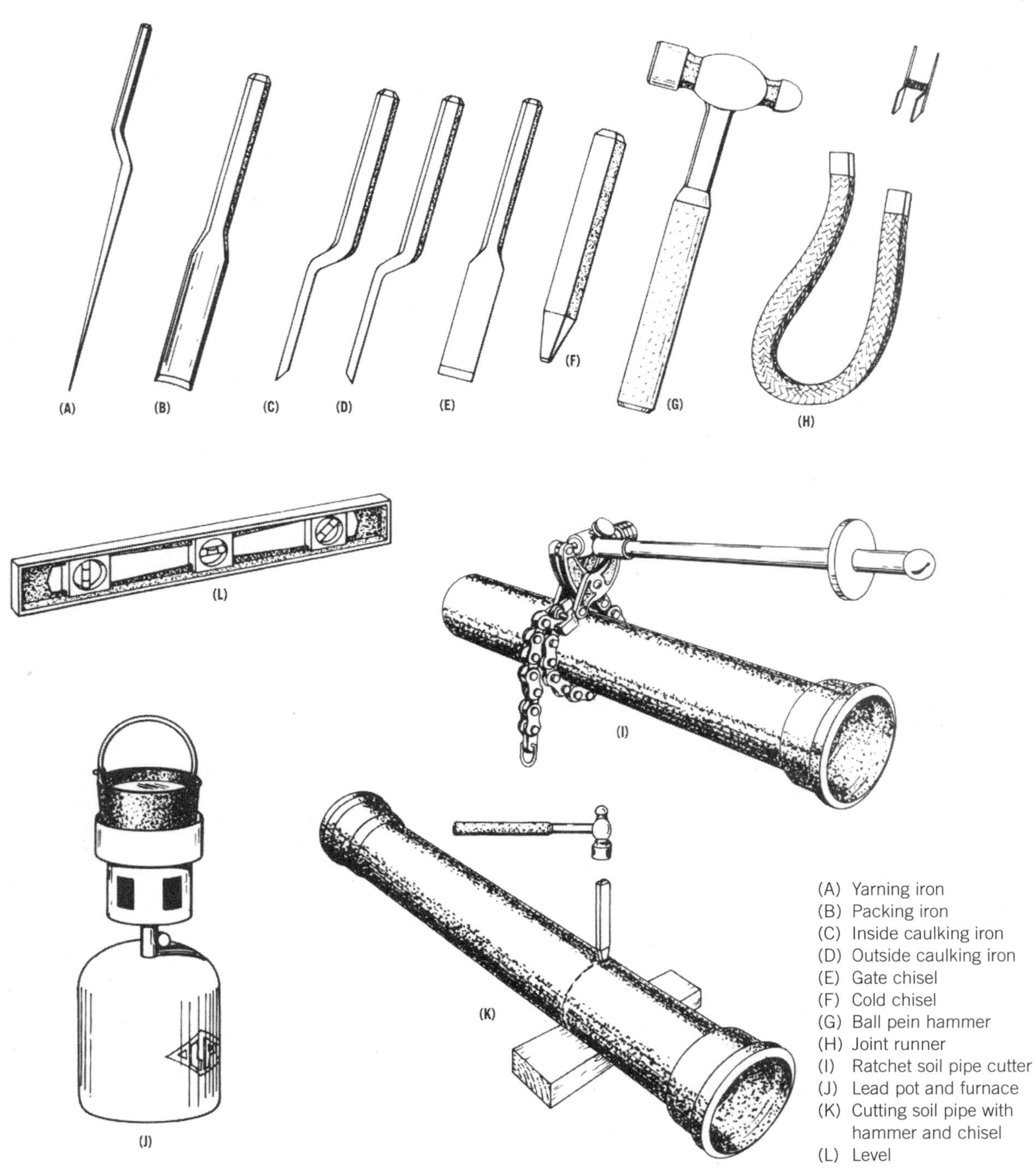

Figure 17–5 Tools used when working with cast-iron soil pipe.

18

Repairing and Replacing Water Heaters

The following paragraphs were taken from the installation and service instructions furnished with a popular brand of water heaters:

> **Gas-Fired Heaters**
> Service or installation of a gas-fired water heater requires ability equal to that of a licensed tradesman in the field. Plumbing, gas supply, air supply, and venting are required.
>
> **Caution:** Natural gas is lighter than air, Liquid Petroleum (LP) gas is heavier than air. If LP gas leaks or escapes it will pocket in a low area of a room or basement. Check local codes and regulations for height of water heater burner above floor line.
>
> **Electric Water Heaters**
> Service or installation of an electric water heater requires ability equivalent to that of a licensed tradesman in the field. Plumbing and electrical work are required.

The information in this chapter is presented as a guide to troubleshooting problems that arise with both gas and electric water heaters. Following the preventive maintenance procedures can avoid costly repairs or replacements.

If a heater tank is leaking, the heater must be replaced. There is no satisfactory repair for a leaking tank. If a relief valve, Figure 18–1, is leaking, working the test lever several times in quick succession to wash out scale, and so on, under the valve seat may stop the leak. If it does not stop the leak the valve must be replaced.

Figure 18–1 A Watts 100XL T & P Relief Valve.
Courtesy Watts Regulator

Gas-Fired Water Heaters

Operation of a gas-fired water heater is controlled by a thermostat and a thermo-couple working together. The thermocouple is made with a bimetal tip, which when heated by the pilot burner to a cherry red, generates an electrical circuit of 7 millivolts or more. This circuit is connected to a magnetic valve in the thermostat. When the thermocouple has delivered the 7 millivolts or more to the thermostat, the magnetic valve is energized, turning on the gas to the main burner. Therefore, we know that for the water heater to heat, the thermocouple must be good. If the pilot light can be ignited while holding down on the PILOT-ON-OFF button, but the pilot flame goes out when the button is turned to ON position, the thermocouple should be replaced.

Following the correct pilot-lighting procedure will help in diagnosing problems with gas-fired water heaters. Figure 18–2 shows the thermostat of an A.O. Smith gas-fired water heater. If your heater is a different brand, the thermostat on your heater may differ in appearance, but it will still have the same basic controls.

The pilot lighting procedure decal should be posted on the heater, or you can use the following:

1. Open the outer jacket door and remove the inner door to the burner chamber.
2. Turn the control knob (PILOT-ON-OFF) located on the thermostat to OFF position.
3. Wait 5 minutes, longer if necessary, for any accumulated gas to escape.
4. Depress the reset button and turn the control knob to PILOT position.

5. Depress and hold down the reset button, light the pilot burner, and observe the thermocouple tip. It should be in the pilot flame, and the tip should be a cherry red.
6. Hold the reset button down for at least 1 minute, and then release it. If the pilot flame stays lit, the thermocouple is good. If the pilot flame goes out, check the thermocouple connection at the thermostat and tighten the connection if necessary. Repeat steps 2 through 6. If the pilot flame will not stay lit, the thermocouple is defective and must be replaced. Do yourself a favor. Call your plumber, who will have a millivolt meter and the tools to fix this problem.
7. If the pilot flame stays lit, replace the inner door and the jacket door. Turn the control knob to ON and set the temperature dial to the desired water temperature. If the pilot flame stays on but the main burner does not come on when the temperature dial is set to the desired temperature, either the water is already at that temperature or there is a problem with the thermostat, possibly energy cutoff (ECO) operation.

The troubleshooting chart in Table 18–1 will help in diagnosing problems with a gas-fired water heater.

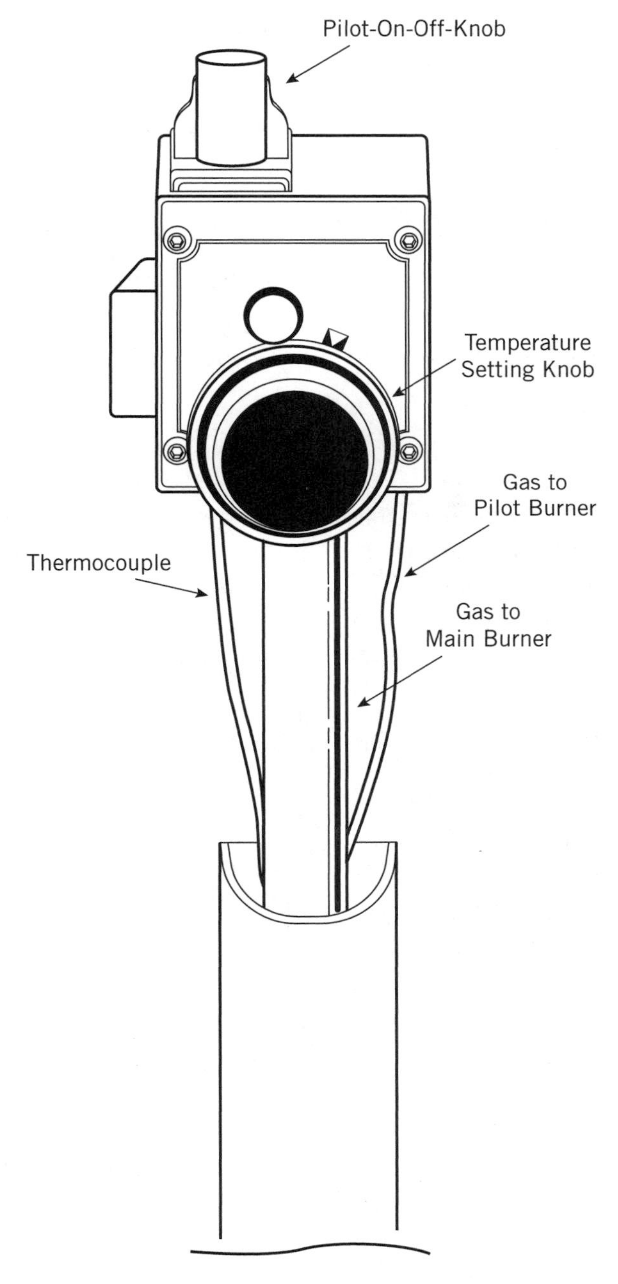

Figure 18–2 Thermostat on A.O Smith gas-fired water heater.
Courtesy A.O. Smith

Energy Cutoff Devices

ECOs are used on water heaters as an additional safety device. If the water temperature in the tank rises above normal operating temperature, the ECO is designed to operate (or open) and cut off the energy supply to the heater. An ECO is incorporated in the thermostat; there are two types, either of which may be used. One type is self-resetting and will reset when the hot water in the tank is replaced by cool water. The other type is a one-time ECO that, when it has operated, requires the replacement of the

Table 18–1 Troubleshooting Gas Water Heaters

Nature of Trouble	Possible Cause	Service Procedure
Pilot will not light.	1. PILOT-ON-OFF knob not correctly positioned.	1. Turn to PILOT position, depress button fully, light pilot.
	2. Pilot orifice clogged.	2. Clean or replace.
	3. Pilot tube pinched or clogged.	3. Clean, repair, or replace.
	4. Air in gas line.	4. Purge air from gas line.
Pilot does not remain burning when reset button is released.	1. Loose thermocouple.	1. Tighten connection at thermostat.
	2. Defective thermocouple.	2. Replace thermocouple.
	3. Defective magnet in thermostat.	3. Call service technician.
	4. Thermostat's one-time ECO has opened.	4. Thermostat must be replaced.
Not enough hot water.	1. Heater is undersized.	1. Reduce rate of hot-water use.
	2. Low gas pressure.	2. Check gas supply pressure and manifold pressure.
Water too hot or not hot enough.	1. Thermostat dial setting too high or too low.	1. Adjust dial setting as required.
	2. Thermostat out of calibration.	2. Thermostat must be replaced.
	3. High water temperature followed by pilot outage.	3. Thermostat out of calibration; replace thermostat.
Yellow flame. Sooting.	1. Scale on top of burner.	1. Shut off heater and remove scale.
	2. Combustion air inlets or flue ways restricted.	2. Remove lint or debris and inspect air inlet opening for restriction.
	3. Not enough combustion or ventilation air supplied to room.	3. Call service technician immediately; turn off gas supply to heater.
	4. Insufficient draft in vent.	4. Call service technician immediately; turn off gas supply to heater.
Rumbling noise.	1. Scale or sediment in tank.	1. Open drain valve, drain 3 or 4 gallons monthly.

Caution: For your safety, *do not* attempt repair of thermostat, burners, or gas piping. Refer repairs to qualified service personnel.

thermostat. A decal on the thermostat should identify which type is used. An ECO rarely operates due to thermostat failure but may operate due to *short draws.* Short draws are short periods of hot water usage (ON-OFF-ON-OFF) repeated frequently over a short time period. This causes the thermostat to turn on every few minutes, resulting in a buildup of very hot water that may trigger the ECO. When the ECO operates under these conditions, the thermostat is not at fault, although if the thermostat has a one-time ECO, the thermostat must be replaced. If a thermostat operated due to short draws, the usage pattern must be changed.

Relief Valve

The Watts 100XL relief valve, Figure 18–1, is a safety device that should be installed on a water heater to prevent injury or damage that could occur due to the excessive pressure and temperature buildup caused by thermostat failure. A relief valve used on a water heater should be American Society of Mechanical Engineers (ASME) rated with a temperature setting of 210°F and a pressure rating of *not more than 150 pounds per square inch.* The discharge piping from a relief valve must extend full size to outside the building wall, to a floor drain, to within 6 inches of the floor, or otherwise as local codes require.

Typical piping connections for gas-fired and electric hot-water heaters are shown in Figure 18–3.

Dip Tube

A dip tube is shown in Figure 18–3. It is needed in a water heater because as water is heated, it becomes lighter and rises to the top of the tank. If cold water came into the top of the tank, it would chill off the hot water. The dip tube takes the cold incoming water to the bottom of the tank where it displaces the hot water going out every time hot water is used. The 1/8-inch hole in the dip tube is to break a vacuum should one form and prevent back-siphonage.

Electric Water Heaters

Most household electric water heaters have two electric elements; lowboy types may have only one. Upper elements rarely burn out, whereas lower elements burn out primarily because mineral sediments, lime, and so on, settle in the bottom of the tank as seen in Figure 18–4 and eventually cover the heating element completely. When a heating element is energized while not submerged in water, it will burn out. When mineral deposits have caused element burn-out, the mineral deposits must be removed before replacing the burnt-out element. Although the cost of the element alone is minor, the labor involved in cleaning out the mineral deposits may come close to the cost of installing a new water heater. To prevent these deposits from accumulating, three or four gallons of water should be drained from the heater each month, through the drain valve at the bottom of the tank.

The troubleshooting chart, Table 18–2, for electric water heaters, will help in determining problems with electric water heaters.

Draining Procedure

Turn off the valve on the cold-water supply to the heater, open the drain valve at the

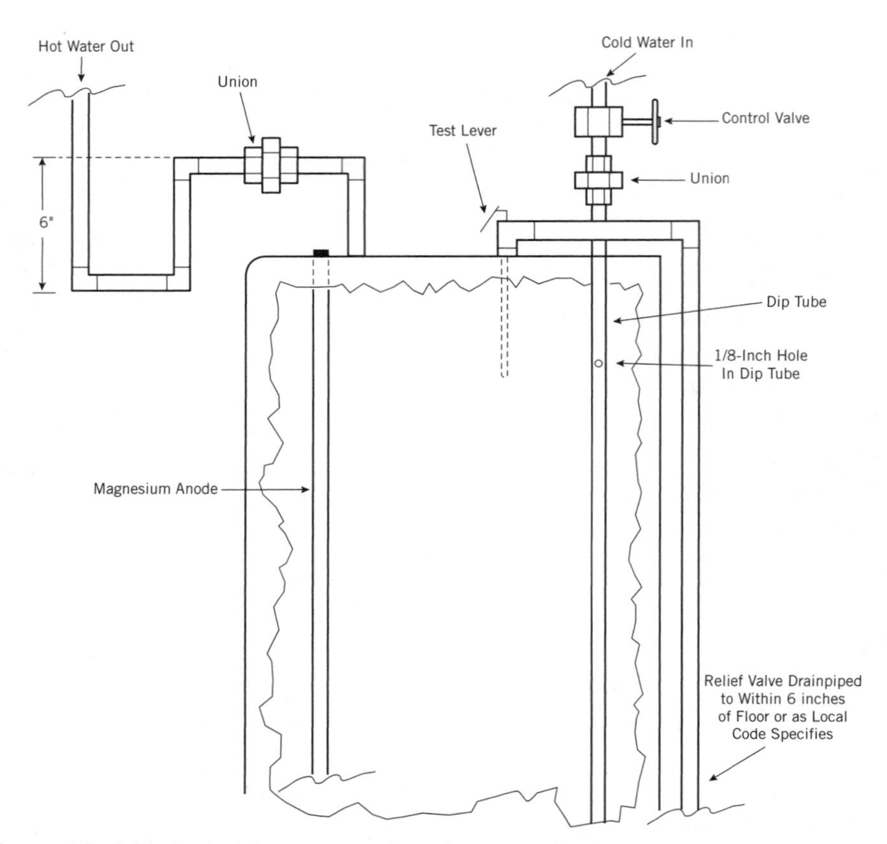

Figure 18–3 Typical piping connections for a gas-fired or electric water heater.

bottom of the heater, and allow water to flow until it is clear. Opening a hot-water faucet prevents a vacuum from forming during the draining process. If the drain valve becomes clogged with sediment, open the cold-water supply valve momentarily; the pressure applied should clear the valve. Plastic drain valves that are installed on many water heaters often leak after use and cannot be shut off. This valve can be removed and a 3/4- by 3-inch brass nipple and a 3/4-inch *ball* valve with a hose adapter on one end can be installed in its place. This valve when open provides a full 3/4-inch opening, and a rod can be inserted through the valve to loosen any sediment in the heater.

The first indication of a burned-out element is when after one shower or one load of clothes is washed, there is no hot water. Figure 18–4 shows a cut-away view of an electric hot-water heater with two elements. Mineral sediments have caused the lower element to burn out. Upper elements, unless defective, never burn out. In fact, upper elements only operate when the tank is filled with cold water. Once the tank is full of hot water, the upper element shuts down and may never operate again. Replacing the element requires the skill and services of a professional. Give yourself a break and call the plumber on this one.

Magnesium Anodes

An anode, a magnesium rod, shown in Figure 18–4, is used in water heaters, both gas and electric, to prolong the life of a glass-lined tank. This is called *cathodic* protection; the anode is slowly consumed (dissolved) as the heater is used. The anode rod should be removed and inspected periodically and replaced when more than 6 inches of core wire is exposed at either end of the rod. The anode is attached to a hexagonal nut on the top of the heater. To remove the anode turn the cold-water inlet valve on the

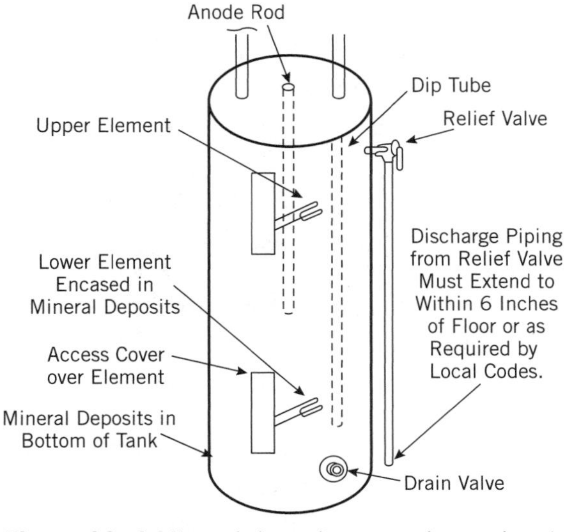

Figure 18–4 Mineral deposits cause burned–out elements.

Table 18–2 Troubleshooting Electric Water Heaters

Nature of Trouble	Possible Cause	Service Procedure
No hot water.	1. Fuses out or blown or circuit breakers switched off.	1. Replace fuses, turn circuit breakers on.
	2. Element burned out.	2. Call service technician.
	3. ECO has opened.	3. Press reset button on thermostat.
	4. Timer has turned electricity off.	4. If special meter for water heater is controlled by timer, recovery is limited to certain hours. If owner's timer has turned electricity off, wait for timer to reset to ON or turn timer manually to ON.
Insufficient hot water.	1. Heater undersized.	1. Reduce rate of water use. If heater is operated by manually adjusted timer, set timer for longer operating periods.
Water is too hot or too cold.	1. Thermostat not set for desired water temperature.	1. Adjust thermostat located under cover plate on jacket of heater.

Caution: For your safety, *do not* attempt to test or replace thermostat or element; refer repairs to qualified service personnel. 240 volts (±10%) are present in electrical panels, timers, thermostats, and elements. User extreme caution when replacing fuses, switching circuit breakers, adjusting timers, or resetting thermostats. *When resetting thermostats, do not remove protective fiber cover* over thermostat.

Note: If pressing the reset button on the thermostat does not result in restoring hot-water supply within a few hours, the element is probably burned out. Call service technician.

heater off, and open a hot-water faucet to relieve the pressure on the line. Then turn the anode nut counterclockwise to loosen it. When it is unscrewed fully, lift the anode out to inspect it. Your local plumbing shop should be able to supply a new anode if you think it is worth the trouble.

When water containing a high percentage of sulfate and/or other mineral content is heated, it can produce a hydrogen sulphide, or rotten egg, odor. Removing the magnesium anode may help this problem, but removal of the anode may also void the warranty on the tank. Water treatment (softening) should minimize the odor problem.

CAUTION: Hydrogen gas can be produced in a hot-water system that has not been used for a period of time, 2 weeks or more. *Hydrogen gas is extremely flammable.* To reduce the risk of injury under these conditions, it is recommended that the hot-water faucet on the kitchen sink be opened for several minutes before using any electrical appliance. If hydrogen is present, it may sound like air escaping when the faucet is opened. There should be no smoking or open flame near the faucet when it is open.

Inspecting the Heater and Vent Piping

The internal flueway of a gas water heater, the vent piping, and draft from a heater should be inspected annually to be certain that they are clean, open, and not rusted out and that a good draft is present. The valve on the gas supply to the heater must be in the OFF position during this inspection. The draft diverter should be removed and the flue baffle lifted out. After inspecting the internal flueway and removing any soot, scale, and so on, reinstall the flue baffle, making certain the baffle is hung securely by its hanger. Inspect the burner chamber and remove soot, lint, or other debris that can interfere with proper combustion. Any soot found indicates improper combustion. The heater should be turned off and left off until qualified technicians can investigate and remedy the problem. The vent piping should be carefully checked for rust spots, holes, or weak spots that would allow combustion products such as carbon monoxide to escape into the house. Carbon monoxide is odorless and deadly. After careful checking, if no problems are found, all parts removed during the inspection must be replaced, and the heater can be placed in service.

One more test should be made while the main burner is on. Check for a good draft in the vent piping by holding a lighted match near the diverter lip as shown in Figure 18–5. If the flame and smoke from the match are drawn up into the diverter, there is a good draft. If the flame or smoke is pushed out and away from the diverter, the draft is poor, allowing combustion products to

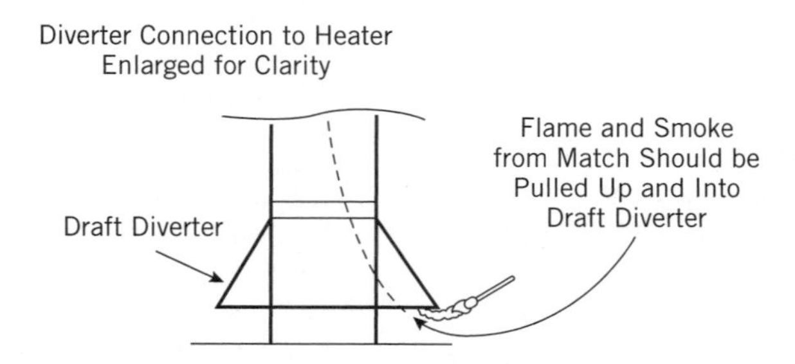

Figure 18–5 Testing a flue for draft.

escape into the area. The heater should be turned off immediately, and a qualified person should be called to correct the problem.

Combustion and Ventilation Air

Proper operation of a gas water heater requires air for combustion and ventilation. If the water heater is installed in an unconfined space within a building of conventional frame, masonry, or metal construction, infiltration air is normally adequate for proper combustion and ventilation. When a gas heater and a gas- or oil-fired furnace are installed in the same room, if the room is tightly closed with little air infiltration, a fresh air inlet may be required.

Figure 18–6 shows a view of a portion of a home containing a fireplace. The illustration shows a gas- or oil-fired water heater and a gas- or oil-fired furnace in the basement. The end result described in this example would be the same if heater, furnace, and fireplace were on the same floor.

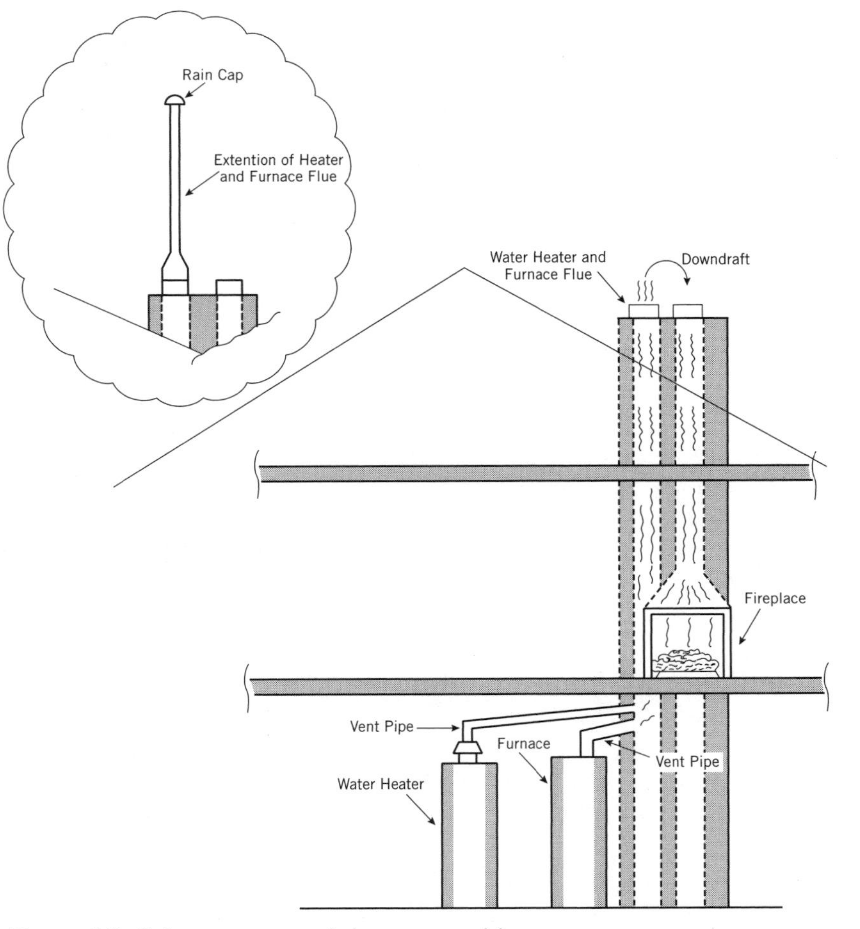

Figure 18–6 A common and dangerous chimney arrangement.

The vent pipes from the heater and the furnace connect to one flue, and the fireplace connects to a second flue.

This is a very common arrangement but a very dangerous one for two reasons: If for any reason raw (unburned) gas should be drawn up the furnace and heater flue, the potential for an explosion exists if the fireplace is being used.

If the fireplace is not being used and the water heater or furnace is on, a downdraft could blow carbon monoxide down the fireplace flue and into the home. And this does happen. When the fireplace is being used, the natural convection (heat rising up the fireplace flue) will normally prevent downdrafts. A gas- or oil-fired water heater or furnace vent pipe or flue should exit the home *not less than 10 feet* from a fireplace flue. If your home has chimneys similar to that shown in Figure 18–6, there is a remedy. The water heater and furnace flue can be extended above the fireplace flue as shown in the inset in Figure 18–6. The extension can be sheet metal, and a rain cap will prevent rain from entering.

19

Caulking in Bathrooms and Kitchens

Water leaks through countertops in kitchens and bathrooms can badly damage plywood or particle board beneath the finished surface. Countertops are usually finished with laminated plastics, (Formica™, and so on) and the plastics are waterproof. But if there is the smallest crack that is not caulked, water can leak through into the subsurface plywood or particle board. When these materials get wet, they expand and push up the finished surface. Usually by the time the damage becomes apparent, the tops are damaged beyond repair. Particle board literally disintegrates when it absorbs water. Figure 19–1 shows areas that should be caulked.

The areas in the kitchen most vulnerable to water leakage are where the countertop meets the backsplash, Figure 19–2. Cracks along these edges may be very small, almost impossible to see, but they are there and water on countertops will find them.

No matter what brand of caulking you buy, the way the caulking is applied is important. The bead should be as narrow as possible and still cover the crack. The tip of the tube should be cut on a 45° angle close to the end. Start at one end of the backsplash and move along steadily. If you think you have missed a spot, go on to the end, and then moisten a fingertip and go back and smooth the bead out, filling the gap.

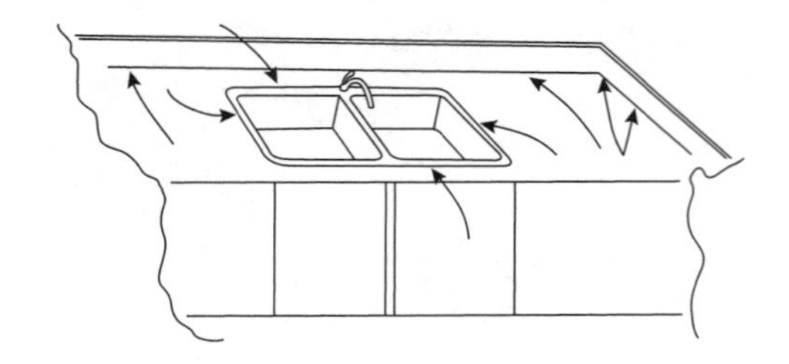

Figure 19–1 Arrows point to areas in kitchen that should be caulked.

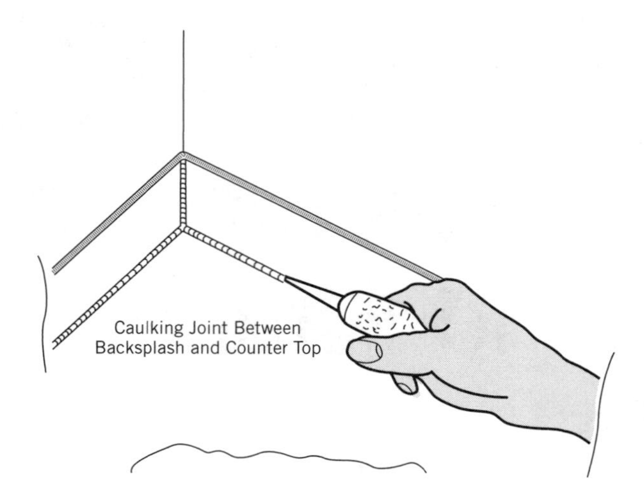

Figure 19–2 Caulking in these areas is especially important.

A common mistake is to think that if a little is good, then a lot is better. Not so. The smaller the bead the better, providing the crack is covered. It is easier to smooth out a small bead than a large one.

If you are going over an area that had been caulked before, it is best to remove the old caulking. This can be done with a thin-edged razor blade knife or box cutter. Hold the cutter tight against the backsplash or tile surface with the blade vertical against the vertical surface and horizontal against the flat surface. This type blade is very sharp and brittle; use it very carefully.

Inspect the edge all around the sink. Although a setting compound probably was used when the sink was set in place, expansion and contraction are constantly taking place in a home. This is especially true in the kitchen, the hottest room in the home.

Some stainless steel sinks are secured to the countertop by a mounting rim with clamps under the countertop. This mounting rim is often not stainless steel. The area under the sink is vulnerable to leaks from the trap, and so on, and this moisture can cause the mounting rim to rust away. The clamp at this point will not hold, and the sink may pull away from the countertop. If this should happen, caulking should be applied at this point until the sink can be replaced.

Figure 19–3 shows the areas where water leakage through joints in tile or around soap dishes, spouts, and escutcheons in a tub could damage the drywall under the tile. It is important for the caulking to seal the area around the spout and the handle escutcheons.

There are many brand names of caulking on store shelves. I've tried them all, acrylic latex, silicone sealants, and even uncured rubber types, but I've found that Polyseamseal All Purpose caulking works best for me. It is white, stays white, and is a water clean-up material.

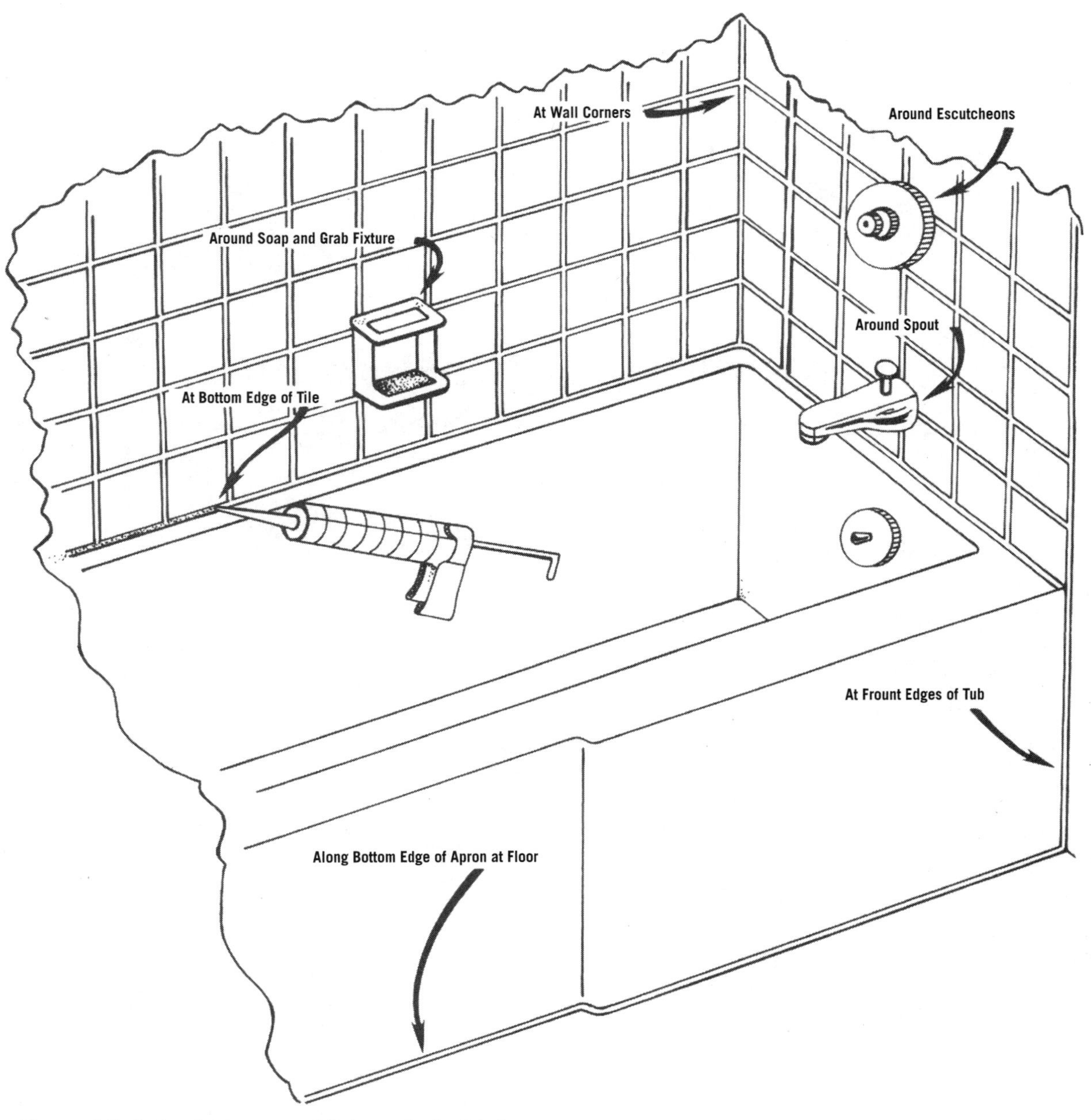

Figure 19–3 Caulking around fixtures in bathtub area.

20

Installing a Circulating Hot-Water Line

Do you go into the bathroom, the one at the far end of the house, turn on the hot-water faucet and wait forever for the hot water to get there? When a gallon or two of water has to be wasted, it's not only a waste of water, it's a waste of energy to heat the water and a waste of your time.

Many of today's homes are ranch-style or trilevel homes where the water heater is at the opposite end of the home from a bathroom. The total developed length of hot-water piping from the water heater to the farthest bathroom may be as much as 30 to 50 feet. When the bathroom is not being used, water in the hot-water pipe is static (not moving) and quickly cools. As the water cools, it gives off heat, adding to the cooling load on the home air conditioning system in the summer. We can solve this problem by insulating the hot-water line to the fixtures and the recirculating line back to the heater.

Before explaining how to remedy this problem, it is important to understand how a water heater works. Cold water entering the heater is conducted to the bottom of the tank through the dip tube, shown in Figure 20–1. Heat is applied at the bottom of the heater, either by a gas flame or electric heating elements. As the cold water is heated, it becomes lighter, or less dense, and rises to the top of the tank. When the hot water reaches the set point of the heater thermostat, the gas burner or the electric element is turned off.

When a hot-water faucet is turned on, hot water is drawn from the top of the heater. Water starts to flow from the faucet and an equal amount of cold water enters the bottom of the heater through the dip tube. When a hot-water faucet in a distant bathroom is turned on, before hot water can reach the faucet, the cold water in the pipe must be drawn off. Unless you are running water in the bathtub, this cold water is wasted, an expensive waste because the cold water is not used and the energy to heat it originally was wasted.

There is a way to get instant hot water at any faucet in your home if the house has a basement or a crawl space and the water piping is accessible. Install a circulating hot-water line.

Figure 20–1 shows two ways in which a circulating line can be installed. Both methods require the installation of a tee in the hot-water piping near the fixture at the farthest point from the water heater. In most cases the piping at this point would be 1/2-inch I.D. copper tubing. The system shown in Figure 20–1 (A) operates on the principle of convection; that is, hot water rises and cooler water falls. For best results the hot-water piping from the heater should slope *up* from the heater to the end of the line. The recirculating line must slope down from its beginning at the tee back to the heater when the system shown in Figure 20–1 (A) is used.

Figure 20–1 (A) and (B) show a 1/2- by 1/2- by 3/8-inch I.D. tee installed near the end of the existing hot-water piping. From this point, a 3/8-inch I.D. line is extended back to the water heater. The connection to the heater is made by removing the 3/4-inch drain connection at the heater and replacing it with the fittings shown in the inset of Figure 20–1 (A). The circulating line is then connected to the tee at the drain connection. Figure 20–1 (B) shows a circulating line that operates using a 1/2-inch 120-volt pump with a built-in timer. The timer can be set to operate only at certain hours of the day. When the system shown in Figure 20–1 (B) is installed, the piping need not be graded. The pump will maintain circulation.

It was pointed out to me that the distant bathroom may only be used once or twice a day. Your local plumber or plumbing supply house handles circulating pumps with built-in timers. Insulating the hot-water piping and installing pumps with built-in timers will result in worthwhile savings over the years.

As I mentioned earlier, the installation of either of these systems is practical only if the home has a basement and the piping is accessible. If you plan on building a new home, I strongly recommend that you install the system shown in Figure 20–1 (B). You will never have to wait for hot water to get to your bathroom again. And incidentally, the cost will be insignificant at the time, and you will regain this cost in a relatively short time.

The connections shown in the inset of Figure 20–1 (A) would also be used if the system shown in Figure 20–1 (B) is installed.

Brass nipples and a brass tee are used at the point where the recirculating line is connected to the heater. If a galvanized tee and nipple were used here, electrolysis would eventually destroy the tee and nipple.

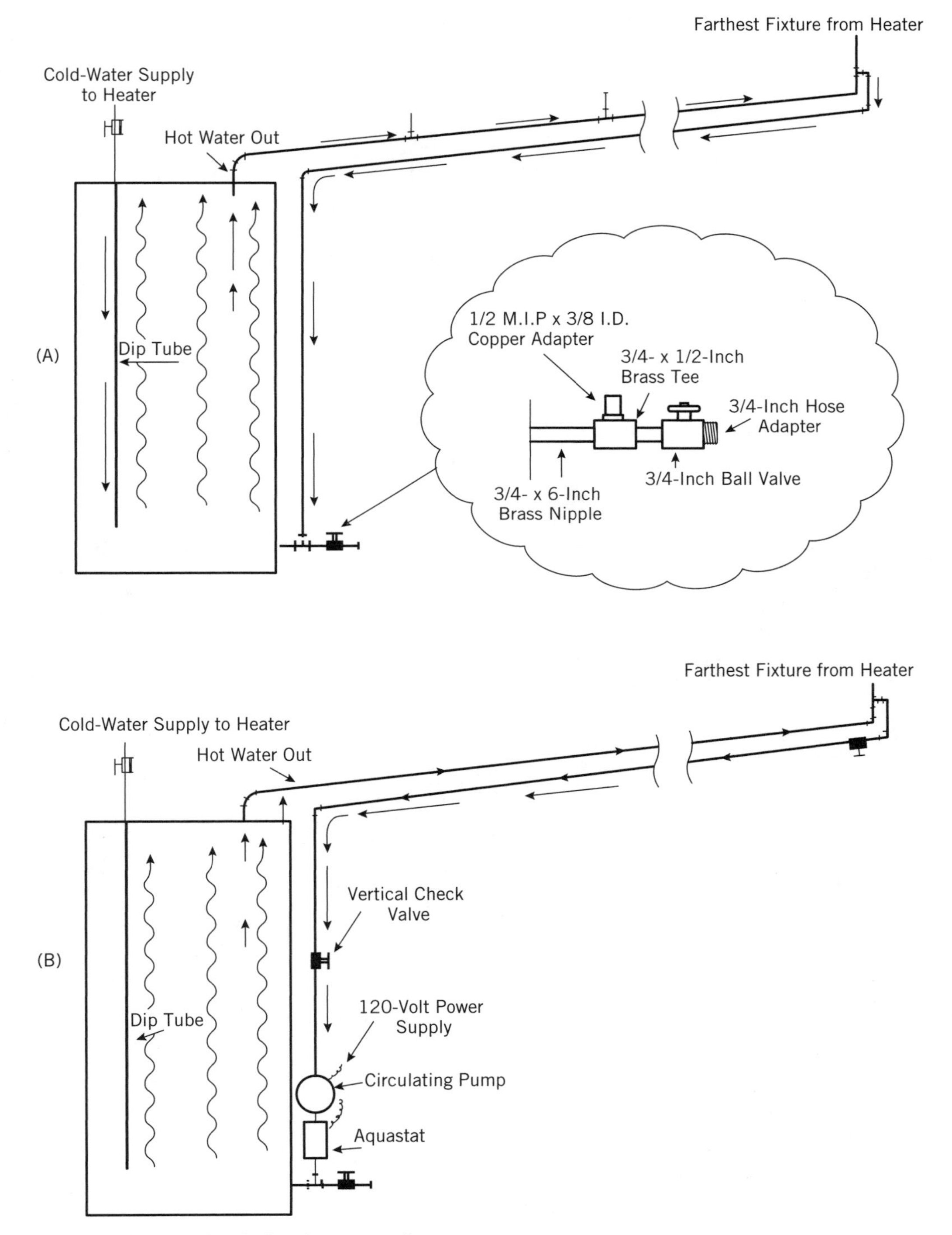

Figure 20–1 Installing a circulating hot-water line.

21

Alternative Ways to Handle Human Waste

As many cities have grown in the last decade, city utilities have been unable to keep up. The trend has been to bigger and bigger homes, situated on larger lots in rural areas. With public utilities unable to reach these areas, developers and home owners have only two solutions for disposing of human wastes: septic tanks or private sewage disposal plants. These private sewage treatment plants, operating in the same basic way as municipal sewage treatment plants, will work in soils and under conditions where septic tanks will not.

Septic Systems

Septic tanks function by a combination of bacterial action and gases. Solids entering the tank drop to the bottom where bacteria and gases cause decomposition to take place. Most of the solids break down into liquids; the insoluble solids settle to the bottom and become sludge. Decomposition in an active tank takes about 24 hours. Bacteria in a septic tank are aerobic bacteria, bacteria that live and thrive in the presence of oxygen. As a part of the breaking down process, a crust forms on the water at the top of the tank.

Figure 21–1 shows a typical septic tank. As water or water-carried wastes enter the tank, an equal volume of effluent leaves the tank at the outlet fitting. The sludge builds up on the bottom of the tank and periodically, depending on usage, the tank must be cleaned or pumped out. In the pumping-out process, the crust at the top of the tank should not be disturbed. The only exception to this rule is if the crust has become coated with grease, thus destroying the bacterial action of the tank. If this should happen, the crust on top will have to be removed. Bacterial action will begin again when the top is placed on the tank and raw sewage enters. Special compounds that are supposed to hasten the resumption of bacterial action can be purchased, but these compounds are rarely if ever needed. The top of the septic tank should be located at a minimum depth of 12 inches below ground level. The actual depth will be governed by the depth of the building drainage piping entering the tank at the tank location.

Sizing the Tank

To work properly, a tank must be large enough to allow decomposition of sewage entering the tank in a 24-hour period. In some areas the size of the tank is based on the number of bedrooms in the home. Most areas now require a minimum size of 1,000 gallons. The septic tank shown in Figure 21–1 is a precast concrete tank, delivered to the job site by a manufacturer and usually set in place when it is delivered. The location of the tank on the job site must conform to local regulations.

Disposing of the Effluent

The liquid discharge (effluent) from the septic tank has not been completely treated. Further treatment will take place in the disposal field by anaerobic bacteria. The disposal field or finger system, shown in Figure 21–1, must be large enough to absorb and/or evaporate the number of gallons of effluent discharged from the septic tank in a 24-hour period.

The absorption rate of the soil where the finger system is located is determined by making percolation tests. The regulations for these tests will vary from county to county and state to state, but in general they will follow these guidelines. Tests are made by digging three holes in the area where the disposal field is to be located. The holes (6 to 8 inches in diameter) should be from 24 to 36 inches in depth, the correct depth for a disposal field. When dug, the holes should be filled to the top with water. When the water has been absorbed into the ground (this simulates the actual working condition of the soil where the disposal field is installed), begin the test. Water should be poured into the holes to a depth of 10 inches from the bottom of the hole. Lay a board across the top of the hole and measure the distance from the bottom of the board to the top of the water. Record this distance and the time. Local health authorities usually have forms for this purpose. Do not add any water during the test. After 1 hour, again measure the distance from the board to the top of the water. The difference between the first measurement and the second is the inches-per-hour absorption rate. Continue the test by recording the distance down to the water level until the test has been conducted for 5 consecutive hours or until all the water has been absorbed. If the water

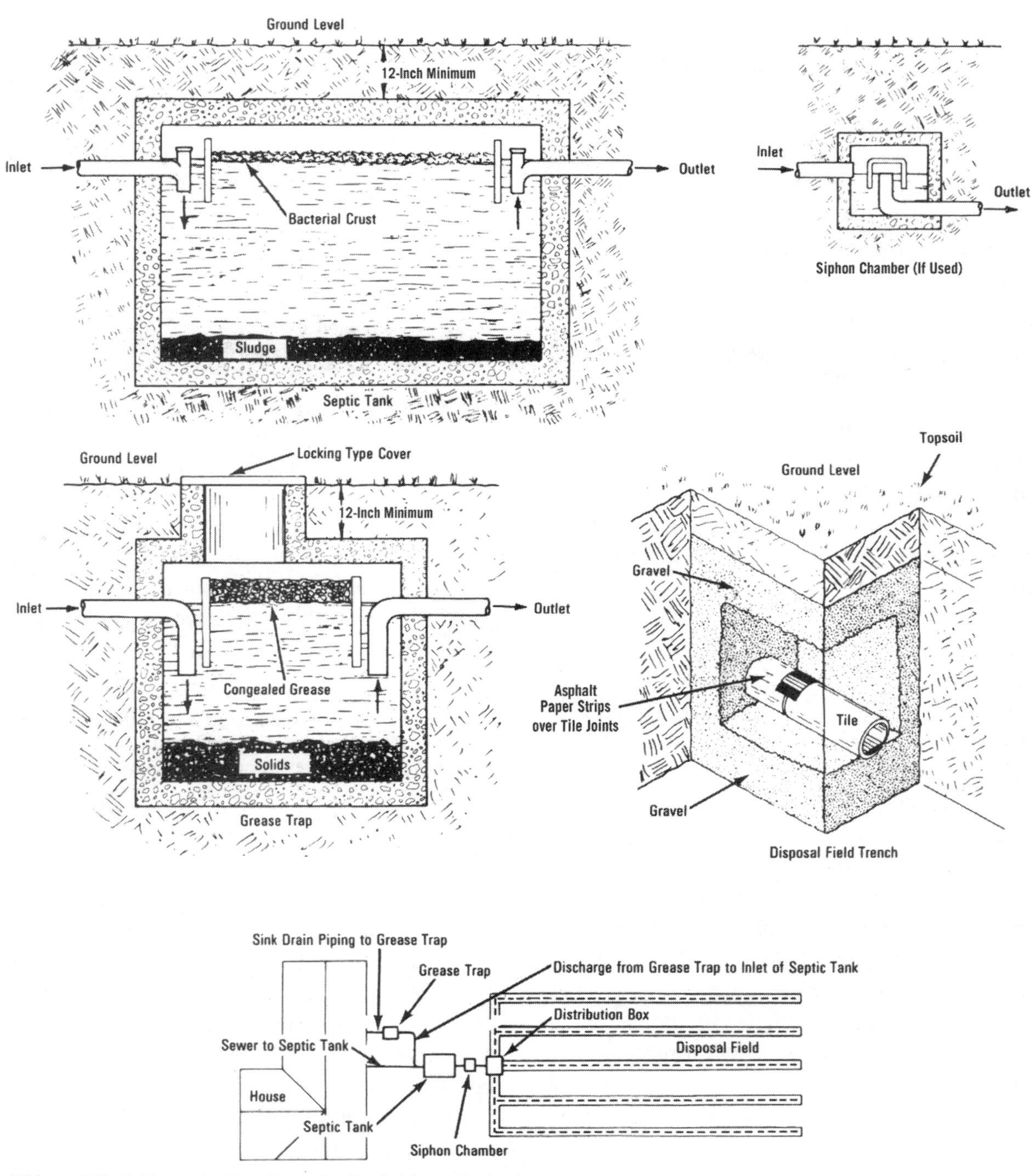

Figure 21–1 The construction of a typical septic tank.

has been absorbed before the end of the 5-hour test period, the test will be ended. Generally speaking, the last hourly rate of absorption is the figure to use in computing the size of the disposal field. If the rate of absorption is less than 1 inch in 60 minutes, a disposal field will probably not work in that area. Areas where clay or hardpan-type soils are near ground level are not satisfactory for disposal fields.

Sizing the Disposal Field

I mentioned earlier that the size of the septic tank is usually determined by the number of bedrooms in the home, with each bedroom equalling two persons. The disposal field must be sized using the rate of absorption established by the percolation tests. If the absorption rate is 1 inch in 60 minutes, the factor 2.35 times the number of gallons entering the septic tank per day will determine how many square feet of trench bottom in the finger system are needed. It is estimated that for every person in a home, 100 gallons will enter the septic tank daily. Using this figure, if the absorption rate is 1 inch in 60 minutes and there are four persons in the home, 2.35 times 400 equals 940 square feet, the square footage of trench bottom needed for the disposal field. If the disposal field has five fingers (trenches), each finger will need 188 square feet of trench bottom. If each trench is 30 inches wide, each trench would be 75 feet long.

If the absorption rate established in the tests was 1 inch in 10 minutes, the factor 0.558 times 100 gallons per day per person would be used. Again using the base of four persons in the household, 0.558 times 400 equals 223.2, which is the number of square feet of trench bottom needed. With five fingers installed in the disposal field, each finger would need 45 square feet of trench bottom. If each trench is 30 inches wide, each finger would be 18 feet long.

The figures used in these examples are from the Indiana State Board of Health recommendations. I selected Indiana because Indiana is more or less in the center of the country and all types of soil are found there. Regulations in other areas may vary somewhat. Regulations in the area of installation must be complied with.

Disposal fields serve two purposes. The fingers of a disposal field provide storage for the discharge from a septic tank until the discharge can be absorbed by the earth. The fingers also serve as a further step in the purification of the discharge through the action of bacteria in the earth. Liquids discharged into the disposal field are disposed of in two ways: by absorption and by evaporation into the air.

Sunlight, heat, and capillary action draw the subsurface moisture to the surface in dry weather. In periods of wet or extremely cold weather, the liquids must be absorbed into the earth. Figure 21–1 shows a siphon chamber installed between the septic tank and the distribution box. The discharge from the septic tank should go into a siphon chamber and be siphoned into the distribution box. The distribution box should direct the discharges so that approximately the same amount of liquid enters each finger.

A siphon (also called a dosing siphon) is desirable because it does not operate until the water level in the siphon chamber reaches a predetermined point. When this level is reached, the siphonage action starts, and a given number of gallons are discharged into the distribution box. The sudden rush of liquid into the distribution box, and then into the separate fingers of the disposal system, ensures that each finger will receive an equal share of the discharge. The discharge thus received by the fingers will have time to be absorbed before another siphon action occurs. A siphon is not essential to the operation of a septic tank and finger system; however, one will improve the efficiency of the system. The lateral distance between trenches in a disposal field will be governed by ground conditions and local regulations.

The construction of a disposal field will vary due to soil conditions. Basically, a typical trench will be as shown in Figure 21–1. A layer of gravel is placed in the bottom of the trench, and field (farm) tile or perforated drain tile is then laid on the gravel. If field tile is used, it should be placed 1/4 inch apart and should slope away from the distribution box at the rate of 4 inches of fall per 100 feet. The space between the tiles should be covered by a strip of heavy asphalt building paper. The trench should then be filled to within 6 inches of ground level with gravel. Some authorities specify pea-sized gravel. Topsoil should then be added to bring the finished trench to ground level. Some settling of the trench will occur; therefore it is wise to mound the earth slightly over the trench to allow for it.

Grease Traps

Grease that is present in dishwashing water can destroy the action of a septic tank and, if allowed to get into the disposal field, will coat the surface of the earth in the field and prevent absorption. Grease is also present in sewage but in small enough amounts that the septic tank can handle it. If the grease from the kitchen can be trapped and removed before it enters the septic tank, the life of the septic tank and the disposal field will be greatly prolonged.

A grease trap must be large enough so that the incoming hot water will be cooled as soon as it reaches the grease trap. A 400-gallon septic tank is ideal for use as a grease trap. The inlet and outlet of the tank have baffles, or fittings turned down, thus the tank is trapped and grease will congeal and rise to the top of the tank where it can be skimmed off. Drain cleaning compounds containing lye (sodium hydroxide) or other strong chemicals should never be used with a septic tank installation. The drainage from a kitchen sink should always go through a grease trap before being connected to the piping entering the septic tank. The top of the grease trap should be within 12 inches of ground level if possible, with a manhole extending up to ground level. A lightweight locking-type manhole cover will provide access when needed to remove congealed grease from the trap. Grease should be removed when it reaches a depth of 3 to 4 inches.

Aerated Sewage Treatment Plants

Residential-sized sewage treatment plants, Figure 21–2, are an excellent alternative to septic tanks. When the soil conditions in an area are such that a septic tank will not work properly, the installation of a residential-sized package sewage treatment plant should be strongly considered. Small, one household size, aeration-type sewage treatment plants that use the same methods as large central plants are now available. These plants are very efficient; when operating properly, the effluent discharge is clear, odorless, and can be chlorinated if necessary to meet health department standards. Local health departments often insist on aeration-type plants instead of septic tanks, especially where the water table is high or where the soil has shown poor percolation. The treatment process, called extended aeration, is a speeded-up version of what happens in nature when a river tumbles through rapids and over waterfalls, purifying itself by capturing oxygen. The JET BAT® plant, Figure 21–2, brings oxygen to the waste water by injecting streams of air into its underground treatment tank and bubbling this air through the waste water.

JET BAT® plants are manufactured in various sizes to meet individual needs as seen in Figure 21–3.

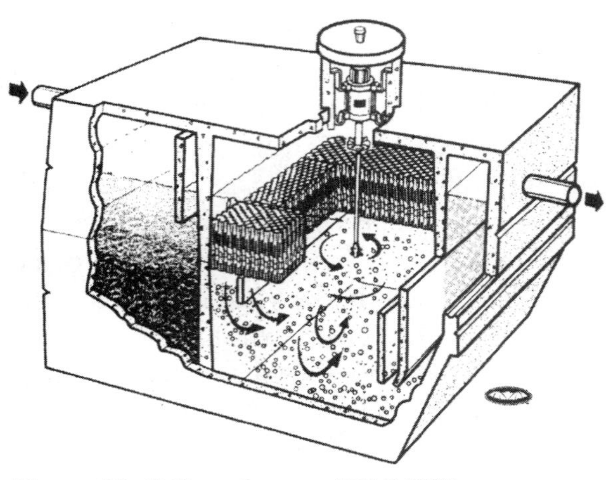

Figure 21–2 One of many JET BAT™ Wastewater treatment plants.
Courtesy JET INC.

How the BAT® Process Works

The compact, efficient Biologically Accelerated Treatment™ Plant (the manufacturer calls it the JET BAT® Plant) has three compartments. The pretreatment compartment, on the left, receives the wastewater and partially treats it physically and biologically before it enters the center treatment compartment.

In the center treatment compartment, technically referred to as a bio-reactor, the JET® Aerator injects fresh air to provide oxygen and mixing to support JET's revolutionary BAT® process. In this process, huge numbers of microorganisms, called a biomass, attach themselves to the submerged JET BAT Process Media®. These microorganisms provide an extraordinarily rapid and high degree of treatment, converting the wastewater to odorless, colorless liquids and gases. Air from the JET® Aerator provides the oxygen required by the microorganisms to complete this process. Mixing ensures that all the wastewater inside this compartment comes in contact with the microorganisms for total treatment.

After treatment the center compartment contents flow into the settling compartment where fine particles settle and return to the treatment compartment. This leaves only a clear, odorless, highly treated liquid for discharge.

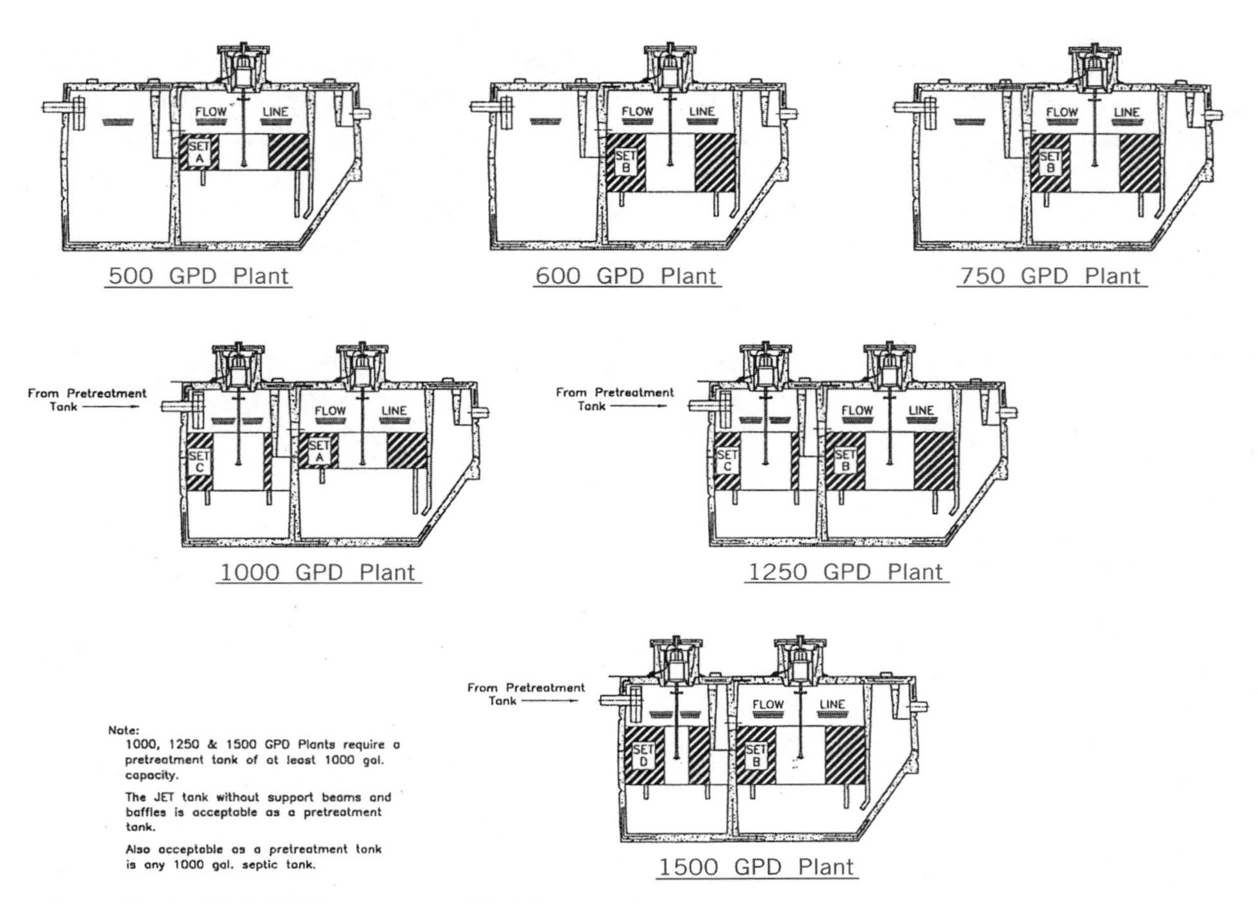

Figure 21–3 JET BAT® Plants are available in several sizes.
Courtesy JET INC.

Septic Tanks versus Treatment Plants

Septic tanks are designed, in order of importance, for sedimentation, sludge storage, and sludge digestion. In conventional tanks all three functions are allowed to take place unseparated in a single compartment, becoming mutually detrimental. The primary treatment applied to the waste stream before discharge to the drain field is physical in nature—sedimentation of solids that can settle.

Septic tanks are a temporary expedient applicable only in areas where soil percolation is sufficient. Bacterial contamination of ground water is probable and drain field failure within the first 3 years can be expected for one-third of the installations.

Now that you know how septic tanks and package sewage treatment plants work, it should be clear that the package sewage treatment plant is an excellent alternative to a septic tank. The effluent leaving a septic tank is only partly digested and must undergo further treatment in a disposal field. If a package sewage treatment plant is maintained and operated properly, the effluent leaving the plant will be a clear, odorless liquid that can be discharged into a stream, if local regulations permit, or into a storm sewer or sand filter bed. As to the relative costs of a complete septic tank and finger system versus a package sewage treatment plant, the difference may be insignificant. Local regulations and plumbing codes must be observed and followed in the areas of work.

22

Private Water Systems

The first question rural residents needing or using a private water system should ask is How much water will I need? Because this is a variable, it is certainly better to err on the heavy side. A good private water system should provide at least 500 gallons per hour to supply the needs for an average family. Using the average consumption per plumbing fixture this figure breaks down as follows:

Activity	No. of Gallons
Flushing a toilet	4
Flushing a bidet	1
Using a shower (varies), average	10
Using a lavatory (washing, shaving, and so on)	2–3
Automatic dishwasher (per load)	15
Cooking (per meal)	5
Automatic clothes washer per load (varies)	30
Yard watering, sprinklers, hoses	100–150 per hour

Naturally, all these figures are variables and depend on the size of the family and usage habits, but a good private water system should provide an ample supply of water at any fixture if other fixtures are being used at the same time.

A good private water system should also provide water in quantity and quality equal to that supplied by a good city water system. The quantity of water delivered by the pump to the pressure tank depends on:

1. The operating pressure of the system, determined by the pressure setting of the pump.
2. The sizing and condition of the piping system. Is the piping size adequate? Is there scale in the piping system, reducing the actual inside diameter of the piping? If well water contains a high concentration of calcium, and so on, and the water has not been conditioned, the piping can be partially or completely blocked by mineral deposits.

The minimum rate of flow from a faucet should be about 3 gallons per minute.

Figure 22–1 shows the various kinds of shallow wells. A shallow well is any source of water where the *low level is not more than 25 feet* below the pump. When water is pumped from a well, the water level will draw down. The lowest level to which it will drop is the level from which it must be pumped.

Figure 22–2 shows three deep wells. A deep well is any source of water where the *low water level* is more than 25 feet below the pump.

Jet Pumps

How a Jet Pump Works

With jet pumps the jet assembly itself forms the suction chamber, and the vacuum is created by the very high velocity of a stream of water passing through the jet. The jet assembly is composed of two parts. The first is a nozzle that creates the high-velocity stream

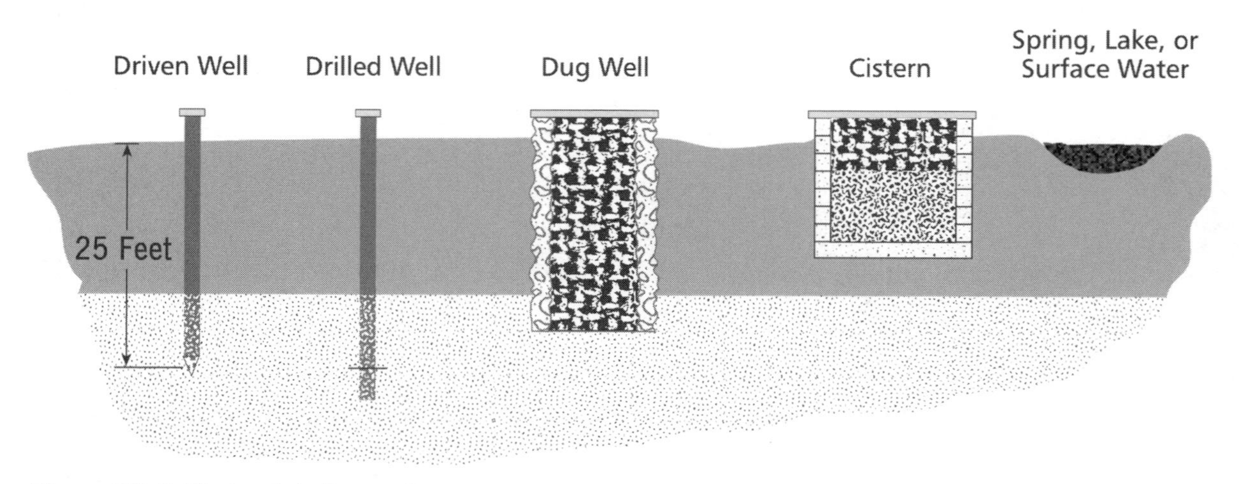

Figure 22–1 Kinds of shallow wells.
Courtesy Goulds Pumps

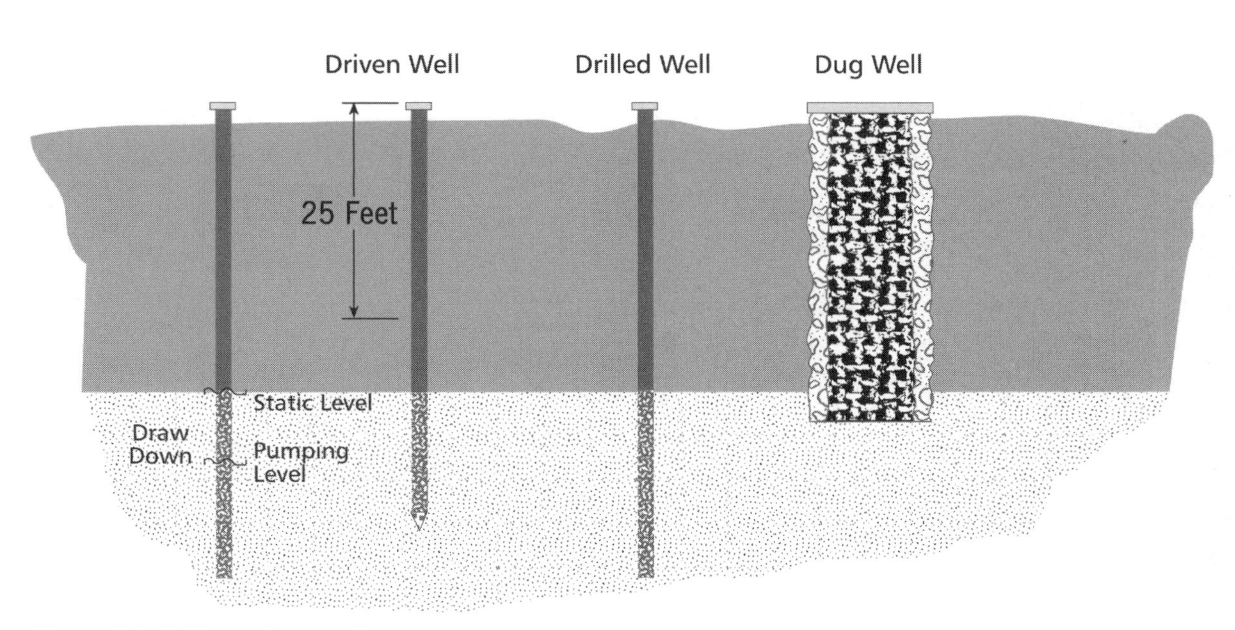

Figure 22–2 Three deep wells.
Courtesy Goulds Pumps

of water, which is injected through a small compartment, the suction chamber, thereby causing the vacuum. Obviously, the suction pipe is connected to this compartment or suction chamber. The vacuum caused by the jet permits the greater pressure of atmosphere on the surface of a body of water to force water into the suction chamber.

The second basic part of the jet assembly is the venturi tube. It is installed in the discharge of the suction chamber. Its function is to convert the velocity of the water into pressure. This is accomplished by the shape of its water passage. Think of a nozzle in reverse. The nozzle speeds up the flow of the drive water, converting pressure into velocity, and when it has passed through the suction chamber, the venturi slows it down again, converting the velocity back into pressure.

Drive water is water that is piped under pressure to the jet assembly or suction chamber. The discharge from the suction chamber or jet assembly is composed of both the drive water and water pumped from the well. The total amount pumped from the well can be used as discharge from the system and is the output, or capacity.

Shallow Well Jet Pumps

Now we know that the operation of the jet system depends on the combined functions of both the jet assembly or suction chamber and the centrifugal pump. These two components of the system are entirely separate, and their locations with respect to each other is a matter of design. A cutaway view of a shallow well pump is shown in Figure 22–3. There is only one pipe extending into

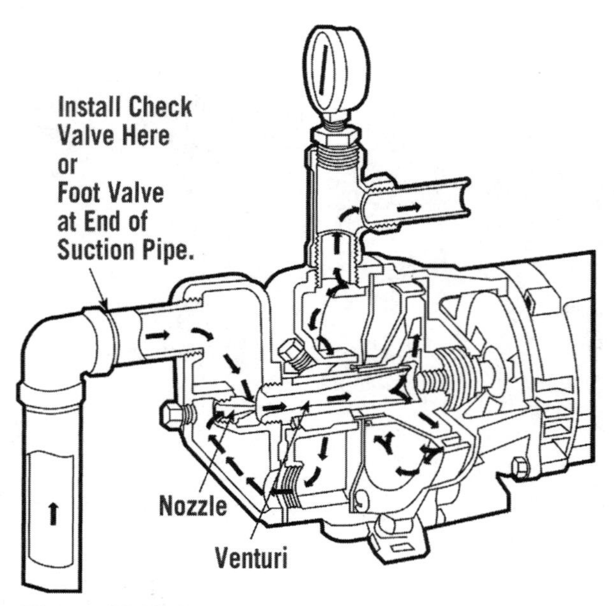

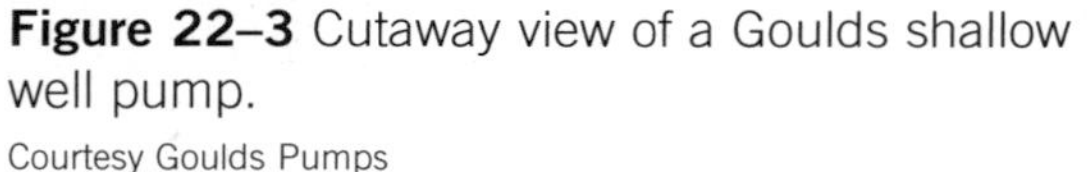
Figure 22–3 Cutaway view of a Goulds shallow well pump.
Courtesy Goulds Pumps

the well or the source of water, the suction pipe. Figure 22–4 shows a diagram of a Goulds shallow well system.

Deep Well Jet Pumps

The only basic or fundamental difference between shallow well and deep well jet pumps is the location of the jet assembly. It must always be located in such a position that the total suction lift between it and the pumping level of the water to be pumped does not exceed that which can be overcome by atmospheric pressure. This means that when this pumping level is at a distance lower than the ground level, which cannot be overcome by atmospheric pressure, the jet assembly must be located at least 5 feet below the low water in the well. A convert-

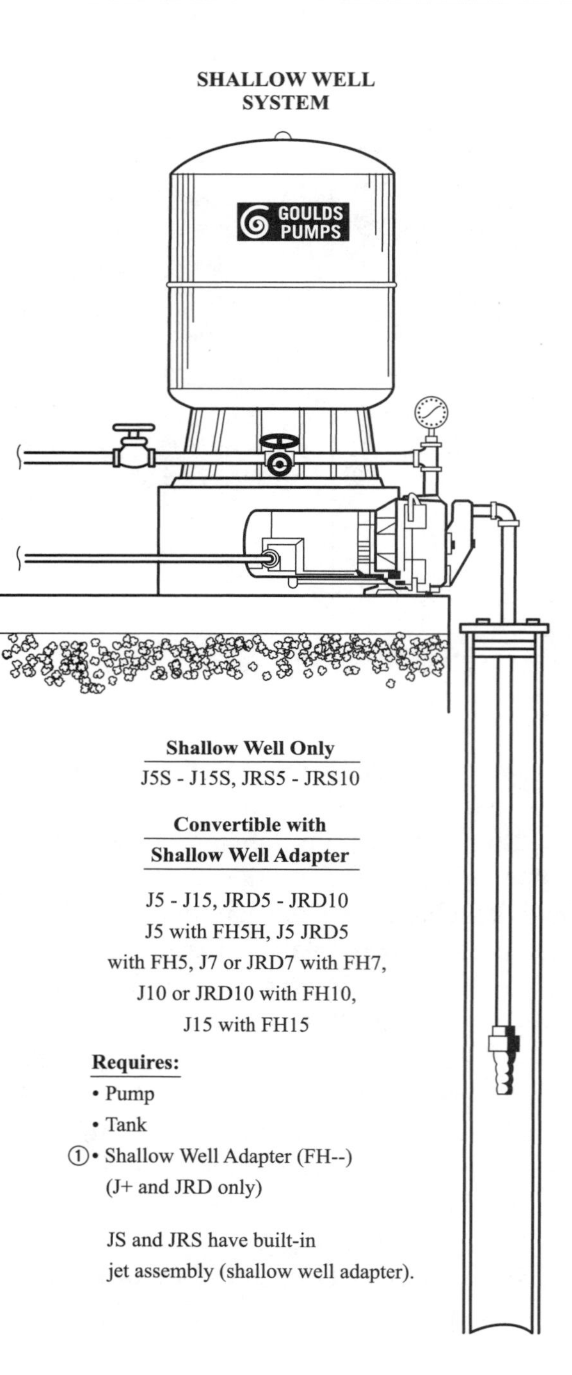

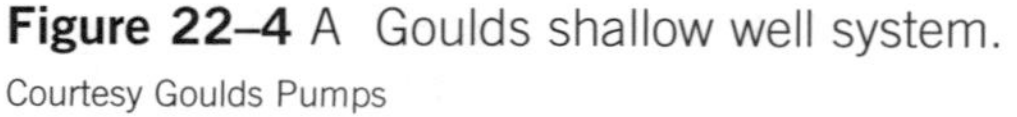
Figure 22–4 A Goulds shallow well system.
Courtesy Goulds Pumps

ible jet pump for shallow or deep wells is shown in Figure 22–5. A diagram of a Goulds twin-pipe deep well system is shown in Figure 22–6.

We must have a closed compartment in which to install the nozzle and the venturi and to form the suction chamber. This part is called the jet body. The jet body may be installed either on the centrifugal pump or on the top of the well casing in a packer system. Its shape is such that it will fit into the casing of a drilled well, and the pipe connections are located for accessibility. There are two on the top side, one for connection to the pressure pipe that supplies the drive water, and the other for connection to the suction pipe that returns both the drive water and the water pumped from the well. For this reason, this connection is one pipe size larger than that for the pressure pipe. Water from the well enters through a third opening, which is on the bottom side of the jet body. A diagram of a Goulds packer deep well system is shown in Figure 22–7.

The last accessory for the jet system is the pressure control valve. It is a valve installed in the discharge piping from the

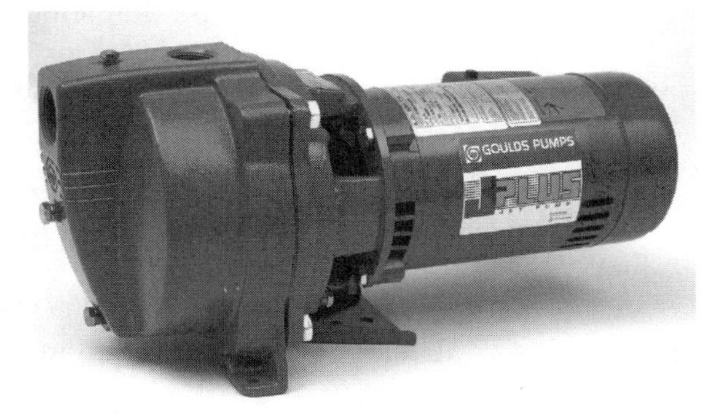

Figure 22–5 Convertible jet pump for shallow wells.
Courtesy Goulds Pumps

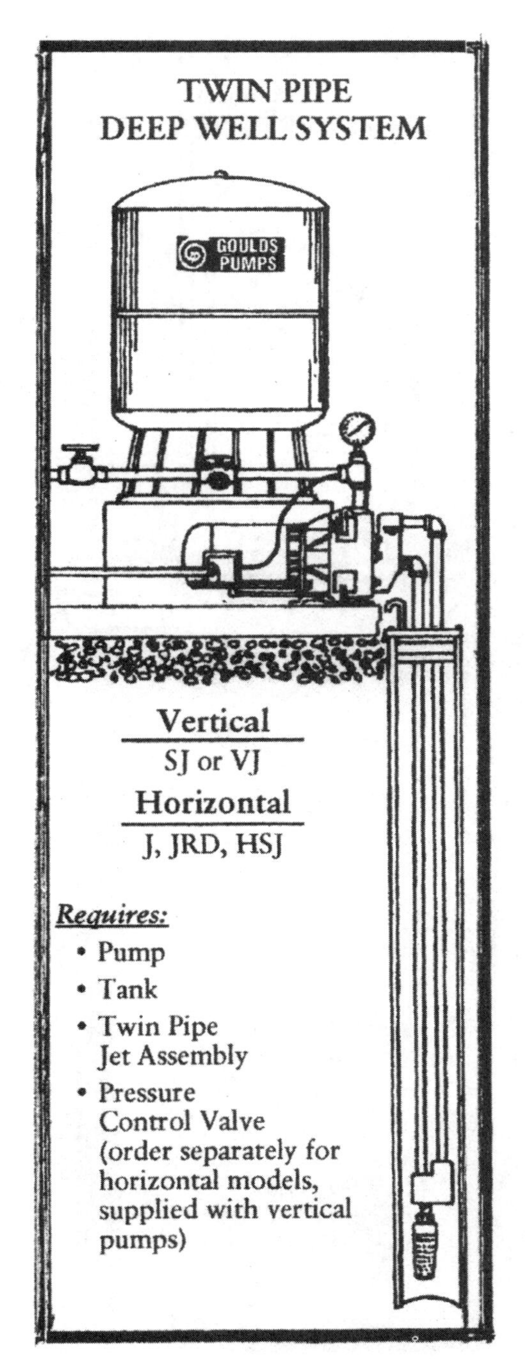

Figure 22–6 The Goulds twin pipe deep well system.
Courtesy Goulds Pumps

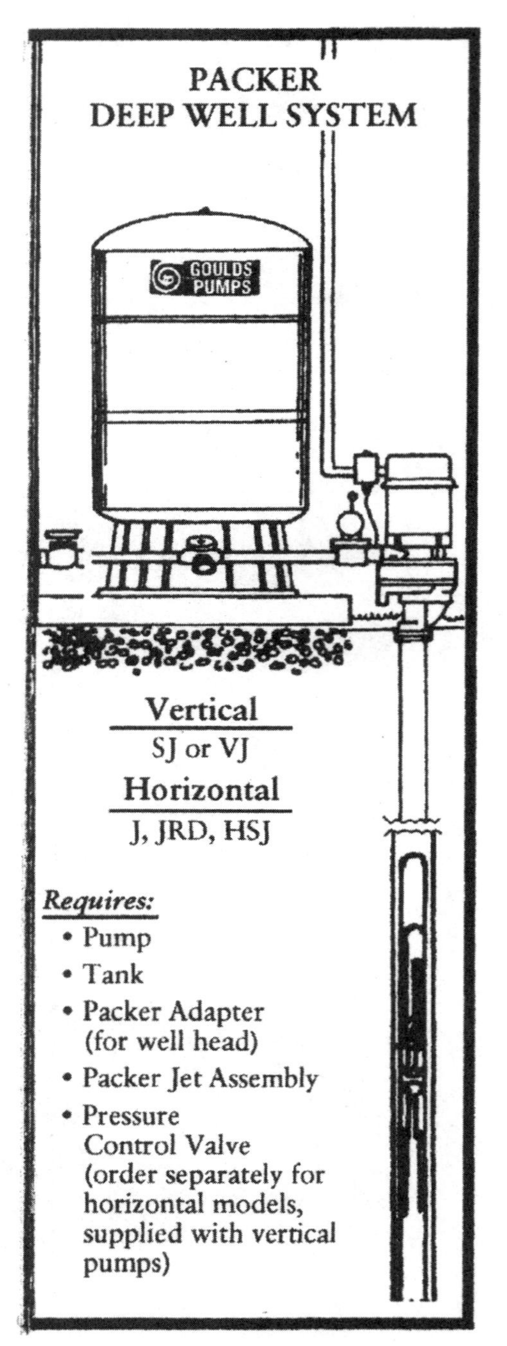

Figure 22–7 Goulds packer deep well system.
Courtesy Goulds Pumps

centrifugal pump between the pump and the tank, in the pump when the pump is mounted on a tank. Used only in deep well systems, its purpose is to assure a minimum operating pressure for the jet.

Submersible Pumps

Submersible pumps, Figure 22–8, are designed for the whole unit; pump and motor, to be operated underwater. A submersible pump does not need to be primed. Once installed and turned on, water flows up the pipe.

The pump end is a multistage (many impellers) centrifugal pump, close-coupled to a submersible electric motor. All of the impellers of the multistage submersible rotate in the same direction by a single shaft. Each impeller sets in a bowl, and the flow from the impeller is directed to the next impeller through a diffuser. These three parts, bowl, impeller, and diffuser, are known as a *stage.* The capacity of a multistage centrifugal (submersible) pump is largely determined by the width of the impeller and diffuser, regardless of the number of stages. The pressure is determined by the diameter of the impeller, the speed at which it rotates, and the number of impellers. The diameter is limited to the size of the drilled well. Most submersible pumps are designed for use in well casings that are 4 or 6 inches or larger in diameter. The diffuser, impeller, bowl, and shaft of a submersible pump are shown in Figure 22–9.

A 1/2-horsepower pump with 7 impellers (designed for capacity) would deliver more water at 80 feet than a 1/2-horsepower pump with 15 impellers (designed for pres-

Figure 22–8 A Goulds submersible pump.
Courtesy Goulds Pumps

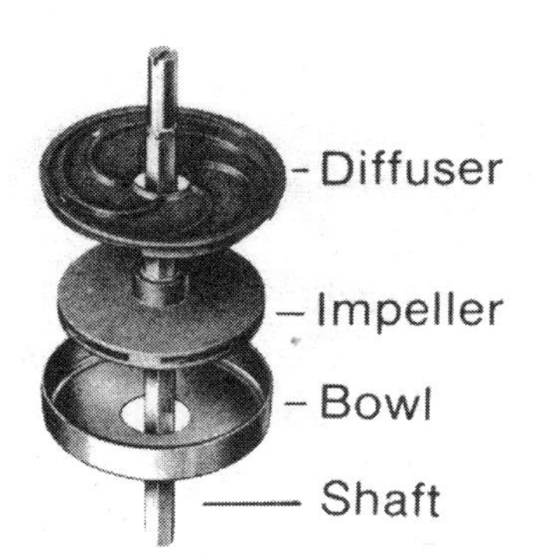

Figure 22–9 One stage of a submersible pump.
Courtesy Goulds Pumps

sure), but the latter pump would be able to raise water from a greater depth.

Well water enters the submersible pump through screened openings at the middle of the unit between the pump and the motor. There is only one pipe connection, which is at the top of the pump. This is the discharge pipe. A check valve is located at the top of the unit to prevent water in the system from draining back when the pump isn't running. Submersible pumps are so much more efficient than jet pumps and the installation so much simpler that a submersible pump should be considered first for all pump applications where the physical dimensions of the source of the water will accommodate the unit in a submerged position. For example, at a 60-foot pumping level and 30 to 50 pounds per square inch pressure a 1/2-horsepower submersible will provide 11 gallons per minute, whereas a 1/2-horsepower jet system will only provide 6.

Centrifugal Pump

The centrifugal pump does two things: It circulates the drive water at the pressure required to produce the necessary velocity in the jet. It also boosts the pressure of the water being pumped from the well, delivering it through the discharge of the system at a satisfactory service pressure. Because the one return pipe from the jet assembly contains both these quantities of water, this return pipe is connected directly to the suction opening of the centrifugal pump. The action of the centrifugal pump can be thought of as that of a paddlewheel. The impeller is a multivane (or blade) wheel, and its design is such that its size, shape, and speed impart sufficient energy to the water in the system to circulate it at the desired rate.

As the water is discharged from the centrifugal pump, it is divided. The drive water, or that amount required to operate the jet, is piped directly to the jet through the pressure pipe. It is continuously recirculated as long as the centrifugal pump is running. The amount pumped from the well is discharged from the centrifugal pump directly into the tank and is the capacity of the system.

Water System Accessories

A complete water system consists of a well, pump, pressure tank, pressure switch, relief valve, and the piping system.

The Pressure Tank

The main purpose of a pressure tank is to balance the capacity of the pump with the demand placed upon it. If the tank is too small, the pump will short cycle (turn off and on too frequently). Short cycling can damage the contacts in the pressure switch or cause damage to the pump motor. A pressure tank should be sized to require the pump to run for at least 1 minute before turning off if the pump motor is rated at less than 1 horsepower. With motors rated at 1 horsepower or more, the pump should run for at least 2 minutes. Pressure tanks are made in various sizes to fit individual needs.

Figure 22–10 shows a Goulds Hydro-Pro pressure tank. The sealed diaphragm-type tank is almost waterloggingproof. For greater system capacity pressure tanks can be installed in series.

The Hydro-Pro™ Water System tank shown in Figure 22–10 will operate at any of the more popular pressure ranges (pump cut-in to pump cut-out). Typically pumps are designed to run in the 20 to 40, 30 to 50, or

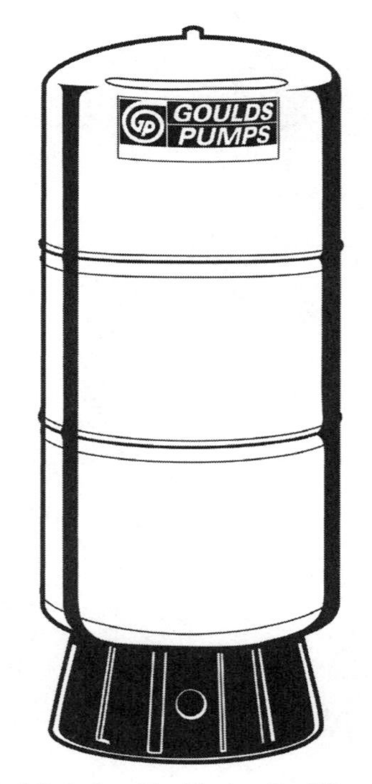

Figure 22–10 A Goulds Hydro-Pro™ pressure tank.
Courtesy Goulds Pumps

40 to 60 pounds per square inch gauge pressure ranges. Although the Hydro-Pro™ Water System tank has been designed to withstand higher pressures, it should not be used in systems exceeding 125 pounds per square inch gauge. Specially designed tanks should be used where higher pressures are encountered.

The Pressure Switch

The operating pressure of a private water system is controlled by a pressure switch, which starts and stops the pump motor at a predetermined setting. A sensing tube connects the switch to some point in the system on the discharge side of the pump. The pressure in the system acts on a diaphragm in the switch, which in turn activates or shuts off the pump. Pressure switches are normally set to turn the pump on at 20 pounds per square inch and off at 40 pounds per square inch. A pressure gauge should be installed between the pump and the tank to monitor pressure.

The Pressure Relief Valve

If the pressure switch should become stuck, pressure in the system could build up and exceed the working pressure of the tank. A relief valve set to open at a pressure 10 pounds above the operating pressure of the tank (20 to 40, 30 to 50, 40 to 60 pounds per square inch) should be installed in the inlet piping to the tank. Goulds tanks are designed for operation on ambient-temperature water systems limited to a maximum working pressure of 100 pounds per square inch gauge. If a water system has the ability to exceed 100 pounds per square inch working pressure, a high-pressure electrical cutoff switch and/or a pressure relief valve rated to open at 90 pounds per square inch must be installed on the system. Failure to install these safety devices can result in personal injury or property damage.

Foot Valve

A foot valve is a combination check valve and strainer used with shallow and deep well jet pumps to prevent water in the piping from dropping back into the well and causing the pump to lose prime.

Waterlogged Pressure Tanks

The most common problem with a private water system is the pump's turning off and on rapidly. This is usually caused by a waterlogged tank. Because water cannot be compressed, an air cushion is needed if pressure is to be maintained in the tank. If there is not sufficient air in the tank, when water is drawn from the tank, the pressure will *drop immediately,* causing the pump to come on. Without sufficient air in the tank, pressure *builds up immediately* and the pump shuts off. The tank shown in Figure 22–11 eliminates this problem. If a galvanized pressure tank *without a diaphragm* is used, the tank should be two-thirds full of water and one-third full of air to maintain a proper air cushion.

CAUTION: There is danger from hazardous voltage. Failure to disconnect and lockout electrical power before attempting any maintenance can cause shock, burns, or death.

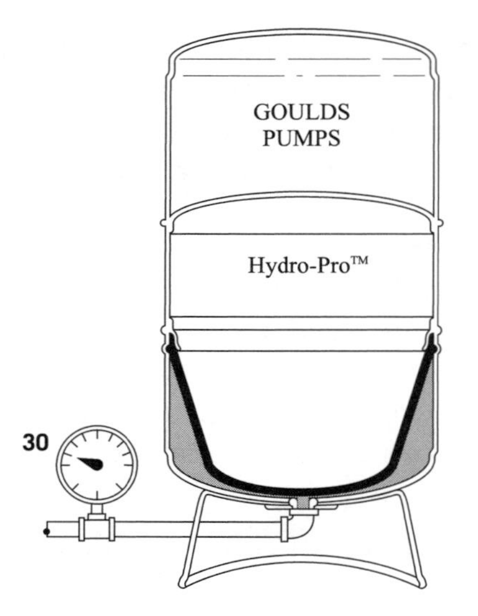

1. Prior to shipping, the tank is pressurized to a standard precharge (28 psig or 18 psig for very small tanks) at the factory.

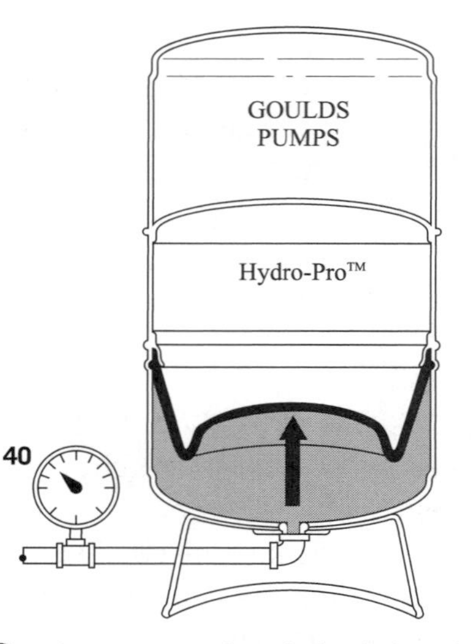

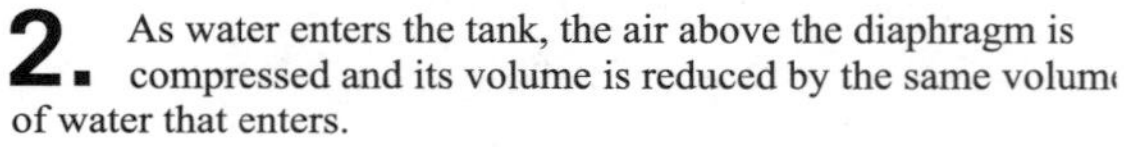
2. As water enters the tank, the air above the diaphragm is compressed and its volume is reduced by the same volum of water that enters.

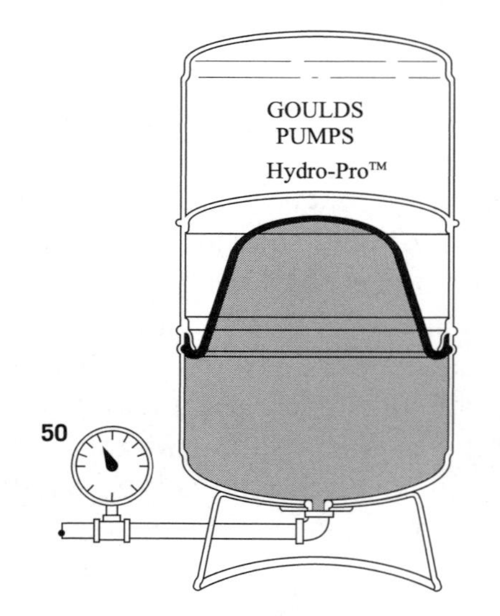

3. The pressure in the tank rises. Water continues to enter until the pump cut-out pressure is reached.The pump shuts off and the tank is now filled.

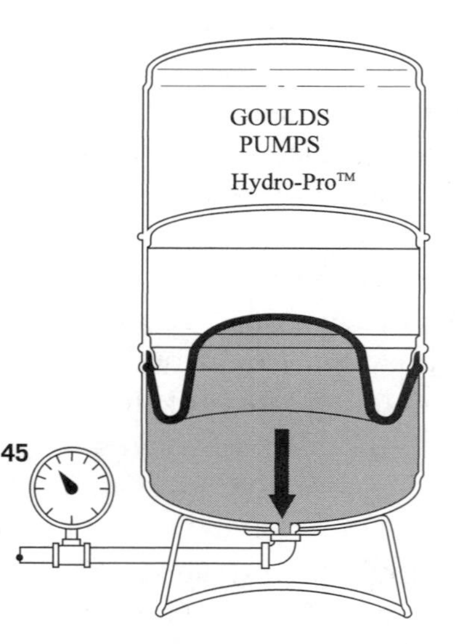

4. Pressure in the air chamber forces water into the system when a demand is made without causing the pump to operate immediately. When the pressure in the air chamber fully drops to the pump cut-in pressure, the pump switch activates the pump and repeats the filling cycle.

Figure 22–11 How a Goulds diaphragm pressure tank works.

Courtesy Goulds Pumps

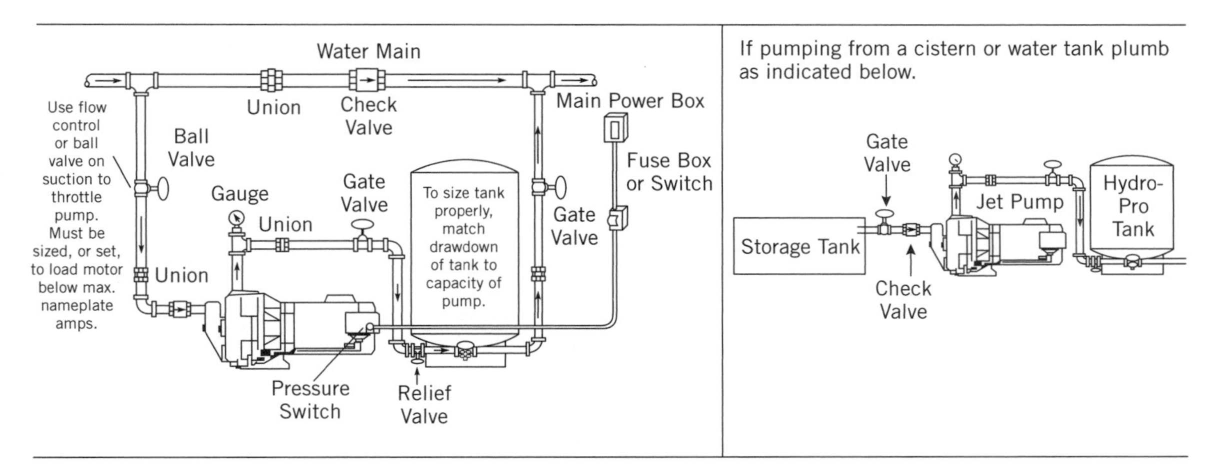

Figure 22–12 Using a Goulds pump as a booster pump.
Courtesy Goulds Pumps

COMMON SYMPTOM AND PROBABLE CAUSE

SYMPTOM

Motor not running. See probable causes 1 through 5.

Little or no liquid delivered by pump. See probable causes 6 through 11.

Pump delivers water but will not shut off. See probable causes 9 through 13.

Pump cycles excessively. See probable causes 14 through 17.

PROBABLE CAUSE

1. Motor thermal protector tripped.
2. Open circuit breaker or blown fuse.
3. Impeller binding.
4. Wiring incorrect; check motor voltage setting.
5. Defective motor.
6. Pump not primed, inadequate suction pipe submergence, or air leak in suction pipe.
7. Discharge or suction plugged, or closed valve(s).
8. Low voltage causing reduced motor speed.
9. Impeller worn or plugged, no jet assembly installed, or jet plugged.
10. System head too high.
11. Suction lift or suction losses excessive.
12. Pressure switch plugged or incorrectly adjusted.
13. Leaks in discharge piping or at house.
14. Defective suction check or foot valve.
15. Waterlogged pressure tank.
16. Pump farther than 5 feet from tank.
17. High friction loss valves between pump and tank (use only fully open ball or gate valves).

23

Drainage and Vent Piping for Island Sinks

Everything concerning a kitchen revolves around the kitchen sink. This has led to the revival of an idea that was very popular several years ago: Place the kitchen sink in the center of the kitchen. If the kitchen area is large enough, this makes good sense. When the plans for a new home or for remodeling a kitchen show the kitchen sink located in an island in the center of the kitchen, it raises a question as to how to properly vent the drainage piping.

To address this problem the Uniform Plumbing Code™ 2000 Edition (IAPMO) came up with the solution: Section 909.0, Special Venting for Island Fixtures. The wording of other plumbing codes may vary slightly from IAPMO's version, but as with many other specific items, the basic concepts of venting this type of fixture installation will agree. Plumbing codes and regulations in effect in the area of work will address the venting of island sinks. The regulations in effect in these areas must be observed.

Figure 23–1 shows one approved method of installing the waste and vent piping for an island sink. Figure 23–2 shows one way to vent an island sink trap. An island kitchen sink is an attractive addition in a large kitchen area.

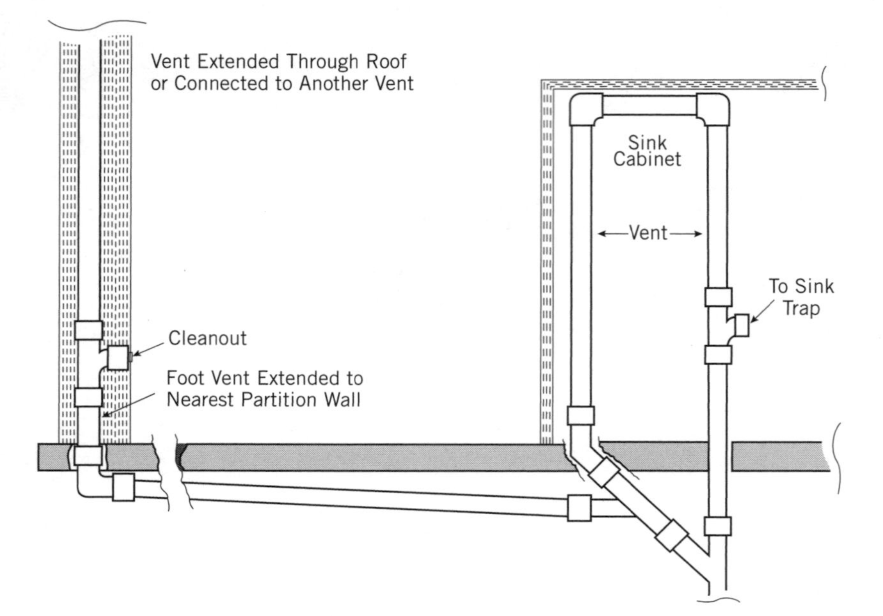

Figure 23–1 Correct method of installing vent for island sink.

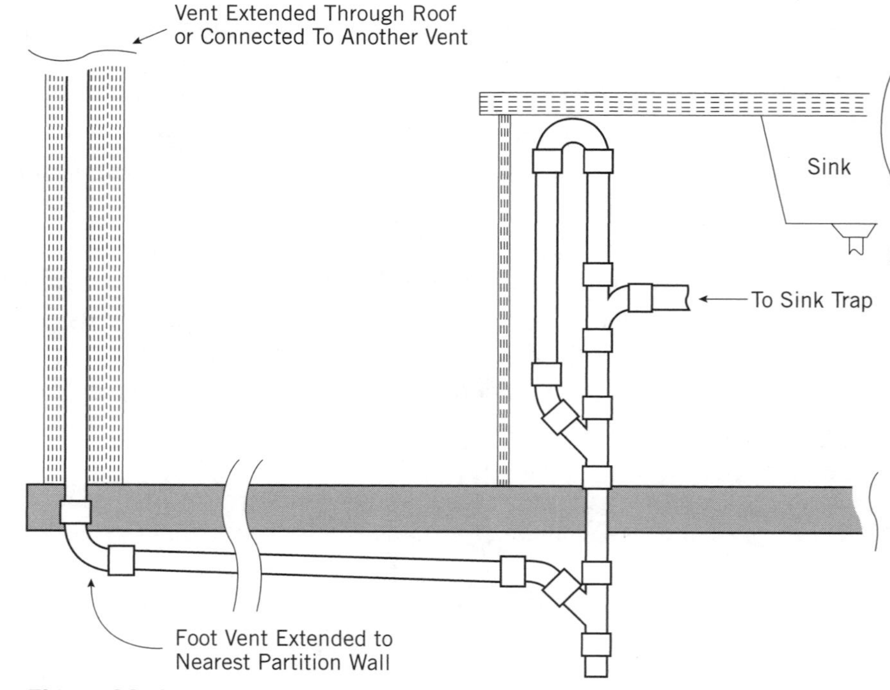

Figure 23–2 The correct way to vent an island sink trap.

24

Plumbing: Ancient and Modern

History

Plumbing gets it's name from the Latin *plumbum* meaning lead. One outstanding example of the Roman's skill in the art of lead working can be seen in the city of Bath, England. The ancient Romans fashioned lead into sheet lead, rolled the sheet lead into pipe, and then piped warm water from artesian wells through these pipes to the famous baths. Lead work and plumbing is one of the oldest building trades. Today, it is a highly skilled trade and one in which any licensed journeyman or master plumber can take pride.

If there is any one thing that is symbolic of modern plumbing, it has to be the toilet. So we'll begin with toilets and we'll end with toilets. The earliest plumbing systems were developed to dispose of human wastes. In the Indus Valley in what are now Pakistan and western India most dwellings had drains for waste disposal as far back as 2500 B.C. Drinking water was supplied by pipes to a royal palace built on the island of Crete around 2000 B.C. This palace also had primitive toilets and a drainage system with air shafts that served as vents. The ancient Romans developed faucets and a sewage system that carried wastes away in rivers and streams.

The quality of plumbing declined after the fall of the Roman empire in 476 A.D. During the Middle Ages in

Europe people disposed of human wastes by throwing them into the streets. Records show that a type of flush toilet was developed in the 1500s. But it did not come into general use because of the lack of plumbing and sewerage systems.

In 1778 an English cabinetmaker named Joseph Bramah patented a flush toilet. It was an innovation that the wealthy classes were quick to enjoy. During the first half of the 1800s toilets became common in England. But most of them drained into pits called cesspools, and these often overflowed. The need for a system to dispose of this human waste was apparent. This led to the invention of septic tanks in the mid-1800s and a modern sewage system began operating in London in the 1860s. Also in the 1860s Sir Thomas Crapper, an English plumber, made improvements in the flush toilet patented by Bramah. Today's flush toilets are the result of Crapper's improvement.

In Italy the Romans had a highly developed plumbing system; water was brought to Rome by aqueducts and distributed to homes. London's first water system, about 1515, consisted partly of the rehabilitated Roman system. The art and science of plumbing has progressed and improved both in Europe and the United States in the past 100 years. Materials used by plumbers have ranged from lead to cast iron to steel, copper, and plastic.

Plumbers are guardians of the public health. The most serious fault in plumbing systems is poor design and/or ignorance of plumbing codes and regulations. The terrible amoebic dysentery epidemic in Chicago during the Century of Progress Exposition in the 1930s was due to a faulty plumbing system in a large hotel. As a result of the poor design of the plumbing system, sewage entered the drinking water of the hotel kitchen. The results of this were catastrophic.

The importance of plumbing codes and regulations cannot be overemphasized. Plumbers hold the public safety in their hands. For this reason many areas today require continuing education as a requirement for holding a license as a journeyman or master plumber.

How We Waste Water

Droughts in various areas of the United States in the late 1990s and into 2000 have brought us face to face with a growing problem. We are running out of potable water. We are wasting water at a catastrophic rate. Toilet flushing is the number one water use in the home, accounting for almost 40 percent of indoor residential water use. Every day 5 billion gallons of water are flushed away, and this is about *3.5 billion gallons more* than is necessary.

Each year approximately 8 million toilets are installed in this country. The American Water Works Association (AWWA) estimates a nationwide savings of 6.5 billion gallons of water per day will be realized by the year 2025 under the existing standards. The introduction of the 1.6-gallon-per-flush toilet, including both reduced tank size and reengineered bowl, has proven to be one of the most efficient water conservation technologies utilized to date.

The Energy Policy Act of 1992 limits the sale of new toilets to those products that use no more than 1.6 gallons per flush. Because of this Act, all new housing units now come equipped with water-saving features, including 1.6-gallon toilets. Many cities from coast to coast now offer financial incentives to home owners to make the switch voluntarily.

Santa Rosa, California, requires that all shower heads and faucets be upgraded at the same time toilets are modified and offers free water-efficient plumbing fixtures in exchange for older models. The city's Go Low Flow plumbing rebate program began in May 1995. At last report, 14,790 toilets have been replaced, resulting in 16,727,116 gallons *per month* in water savings.

Customer Satisfaction Levels

When water-saver toilets first made their appearance, many customers were dissatisfied. The common complaint was that two, three, or even more flushes were needed to empty the bowl. That has changed. Companies changed their designs to ensure that their products met government

Figure 24-1 The Peerless Pottery Madison 1.6-gallon-per-flush toilet.
Courtesy Peerless Pottery

Figure 24-2 The Peerless Pottery McKinley 1.6-gallon-per-flush handicap toilet.
Courtesy Peerless Pottery

specifications. Performance surveys show high levels of customer satisfaction for home owners. Surveys in Santa Rosa, California, show that about 95 percent of consumers who participated in the city's Go Low Flow program find the new products *as good or better* than the replaced models.

The Peerless Pottery Madison model round front water closet shown in Figure 24–1 uses no pumps, grinders, pressure tanks, or other complicated apparatus. It is available in 10-, 12-, and 14-inch rough-in models. This newly developed toilet will provide a positive flush with complete and thorough bowl cleansing while using less than 1.6 gallons per flush of water. The Controlled Rate Flapper used in Peerless Pottery toilets is foolproof. It works every time.

The Peerless Pottery McKinley model 17 1/2-inch-high handicap 1.6-gallon-per-flush toilet is ideal for the elderly and physically challenged and meets all requirements for a handicap installation. The elongated bowl toilet shown in Figure 24–2 is available in 10-, 12- and 14-inch rough-in models to fit existing plumbing.

So now we have come full circle. We started this chapter explaining the origin of the flush toilet and we finish by showing the latest advancements in water-saving toilets.

NOTE: For any brand of water-saver toilet to work properly, the flapper-type tank ball designed for that toilet must be used when replacing this part.

Appendix

Fittings

It helps to know the name of the fitting you need. Some of the most common steel or malleable pipe fittings are shown in Figure A–1. The names are the same whether the fittings are steel, cast iron, or copper. Elbows and tees are made in straight pipe sizes and also in reducing patterns. For example, a 1/2-inch elbow has 1/2-inch threads in each opening. A 1/2- by 3/8-inch elbow has 1/2-inch threads in one opening and 3/8-inch threads in the other.

Tees are also available either as straight pipe sizes or in reducing patterns. Examples are 1/2 by 1/2 by 1/2 (straight pipe size) and 1/2 by 1/2 by 3/8 (reducing tee).

A tee is always "read" as end, end, side; thus a tee that is 1/2 inch on one end, 1/2 inch on the other end, and 3/8 inch on the side is a 1/2- by 1/2- by 3/8-inch tee.

Couplings are often confused with unions. Couplings join pieces of pipe together, as do unions, but unions must be used when joining two pieces of pipe together between pipes or fittings that cannot be turned or moved.

Pipe nipples are short lengths of threaded pipe, from "close," or all thread, to 12-inch length. "Street" fittings are fittings that are made with male threads on one end and female threads on the other, or, in the case of copper sweat

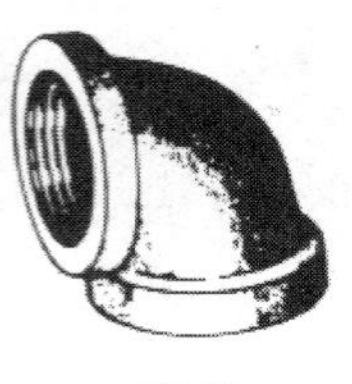

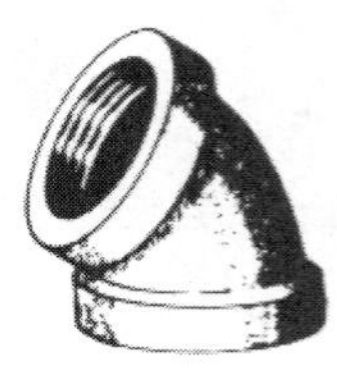

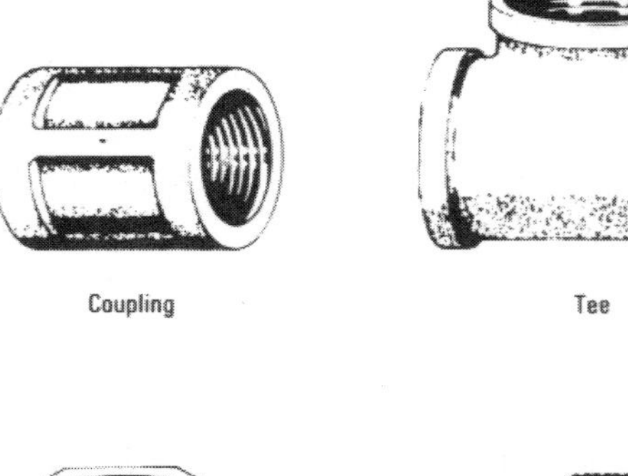

Figure A–1 Various pipe fittings.

fittings and PVC or CPVC fittings, male pipe size on one end and a female socket on the other end. Threaded street fittings are seldom used; the male end is restricted and street fittings are difficult to grasp with a pipe wrench. A 90° elbow and close nipple or a 45° elbow and close nipple are better than 90° or 45° street elbows.

Traps on fixtures are not there for the purpose of stopping or catching any object entering the drain of a fixture except in one case. Certain hospital-type fixtures or fixtures used in connection with the making of plaster casts have a plaster-catching trap. Chemistry labs also have plaster-catching traps. The traps on plumbing fixtures in the home are there to prevent sewer gas from entering the home through the fixture drain piping.

Condensation

When the warm moist air in a bathroom comes in contact with the cold surface of the toilet tank, the moisture in the air is condensed and forms as water droplets on the surface of the tank. A thermostatic supply valve that mixes hot and cold water to supply the toilet tank is made to solve this problem.

Thawing Frozen Piping

If frozen water pipes are in an accessible location, they can be safely thawed by pouring hot water on the frozen pipe or by wrapping rags dipped in hot water around the frozen areas. A heat lamp can also be used to thaw frozen pipes if the lamp is kept a safe distance away from combustible materials. An electric hair dryer aimed at the frozen areas will also thaw the pipe, or an electric heating tape wrapped around the frozen areas can be used. If the heat tape is placed on the pipe and plugged in to an electrical outlet when freezing weather is expected, it should prevent the pipe from freezing.

If the correct voltage and current settings are used, arc welders can be used to thaw buried frozen water services.

If freezing temperatures are imminent, open faucets and let the faucet drip. Moving water will not freeze.

Noise in the Piping System

A hammering or crackling noise in a gas-fired water heater is caused by lime or other minerals that solidify on the bottom or sides of a water heater when the burner is on. This mineral coating is not watertight and small droplets of water are trapped under this coating. The next time the burner comes on, the trapped water is heated and expands. This expansion of the water trapped under the coating causes the coating to flake off, and in the process the rumbling, hammering, or crackling noise is produced. This noise is not dangerous; it's just a nuisance.

If you have washer-type faucets, you may have hammering noises when a faucet is turned on. This noise is generally only heard when the faucet is only *slightly* opened, with a very small stream flowing. The noise then will be a slow hammering noise that will increase in volume and intensity as the faucet is opened wider. At wide open position the noise usually disappears. This noise is caused by a loose bibb (faucet) washer. Remove the stem and tighten the screw holding the bibb washer in place. The noise will be gone.

A whistling-type noise is often heard from a toilet tank. This noise is caused when an old-fashioned type ballcock starts to shut off the water flow as the proper water level in the tank is reached. The rate of water flow through a ballcock varies. Water flows fast as the tank starts to fill, but as the float ball rises with the water level and in turn pinches off the water flow, a whistling noise is often produced. The installation of a fill valve of the type shown in Chapter 6 will eliminate this source of noise.

Noise in the piping system is also caused by the sudden closing of a valve. This causes a shock wave to travel through the piping. When the wave reaches the end of the pipe, it reverses, and the hammer-blow noise is produced. This noise is very common when a dishwasher or a clothes washer is being used. These appliances use electric solenoid valves. When they close, they close very quickly, creating a shock wave in the pipe. Installation of an air cushion (shock absorber) on hot- and cold-water piping will cure this problem.

General Information

Every home owner should know where the main shutoff valve is located on the water supply to the home. Knowing where this valve is located and how to operate it may prevent costly damage. Follow the cold-water connection from the water heater location back to the point where the water pipe enters the house. The valve should be located at or near the outside wall. In some areas if the house is built on a crawl space, the main valve could be in a water meter pit in the yard. The main cold-water valve controls all the water to the home. If it is necessary to turn off only the domestic hot water (not hot-water heat) to the house, the shutoff valve on the *cold*-water supply to the water heater will turn off the hot water.

An emergency repair can be made by using one or two adjustable radiator or heater hose clamps and a small piece of inner tubing. If the clamps are placed over the hole and tightened, they should hold pressure until a permanent repair can be made.

Water backing up through a floor drain can be prevented by installing a 2-inch backwater valve in the floor drain. Most basement floor drains are made with a 2-inch tapping in the body of the drain, under the lid. The backwater valve is a fitting with a rubber seat on the bottom of the fitting and a metal or plastic float ball. The float ball will drop to permit the passage of water through the inlet side of the floor drain, but the ball will rise and seat itself against the rubber seat of the fitting, preventing water from backing up through the floor drain.

The word *vent* when applied to a plumbing system is a pipe that provides a flow of air to and from the plumbing drainage piping. Proper venting is necessary to permit good drainage flow and prevent siphonage in a drainage system.

To eliminate the possibility of accidental scalding, the thermostat on a water heater should be set to deliver a maximum hot-water temperature of 120°F. The hot-water temperature should be checked using a thermometer at the water heater or at a hot-water faucet nearest to the water heater.

Bathtubs now in use that do not have a slip-resistant bottom can be made safer by an application of nonslip tape.

How to Figure Capacities of Round Tanks

There are times when it is necessary or desirable to be able to compute the contents in gallons of round tanks, cisterns, or wells. A tank 12 inches in diameter and 5 feet in length will hold 29.3760 gallons of water. A well casing 4 inches in diameter with 10 feet (120 inches) of water standing in the casing has 6.528 gallons of water standing in the casing.

Shown are the two old standard methods of figuring the contents of round tanks. Also shown are two shorter and easier methods. The shorter methods are not only easier, but there is less chance for error because fewer steps are used in obtaining the results.

C = capacity in gallons
D = diameter
L = length
0.7854 = area of a circle
231 = cubic inches in gallons
7.48 = gallons in cubic foot

The old standard methods: When measurements are in inches:

$$\frac{D^2 \times 0.7854 \times L}{231} = \text{capacity in gallons}$$

When measurements are in feet:

$$D^2 \times 0.7854 \times L \times 7.48 = \text{capacity in gallons}$$

Problem

A tank is 12 inches in diameter and 5 feet in length. How many gallons will the tank hold?

Using method 1:

1.

$$\begin{array}{r} 12 \\ \underline{\times\ 12} \\ 144 \end{array}$$

2.

$$\begin{array}{r} 144 \\ \underline{\times\ .0754} \\ 576 \\ 720 \\ 1152 \\ \underline{1008} \\ 113.0976 \end{array}$$

3.

$$\begin{array}{r} 113.0976 \\ \underline{\times\ 60} \\ 6785.8560 \end{array}$$

4.

$$\begin{array}{r} 29.3760 \\ 231 \overline{\big) 6785.8560} \\ \underline{462} \\ 2165 \\ \underline{2079} \\ 868 \\ \underline{693} \\ 1755 \\ \underline{1617} \\ 1386 \\ \underline{1386} \end{array}$$

Answer: 29.3760 gallons

Method 2 also requires four steps to arrive at the answer:

1.

$$\begin{array}{r} 1 \\ \underline{\times\ 1} \\ 1 \end{array}$$

2.

$$\begin{array}{r} 0.7854 \\ \underline{\times\ 1} \\ 0.7854 \end{array}$$

3.

$$\begin{array}{r} 0.7854 \\ \underline{\times\ 5} \\ 0.39270 \end{array}$$

4.

$$\begin{array}{r} 3.927 \\ \underline{\times\ 7.48} \\ 31416 \\ 15708 \\ \underline{27489} \\ 29.37396 \end{array}$$

Answer: 29.37396

The answer does not agree with the answer shown in method 1. The difference in the answers is minor, however.

The shorter methods in figuring tank capacities are:

Using method 3:

$$C = D^2 \times L \times 0.0408$$

When measurements are in feet and inches:

1.

$$\begin{array}{r} 12 \\ \underline{\times\ 12} \\ 144 \end{array}$$

2.

$$\begin{array}{r} 144 \\ \underline{\times\ 5} \\ 720 \end{array}$$

3.

$$\begin{array}{r} 720 \\ \underline{\times\ 0.0408} \\ 5760 \\ \underline{28800} \end{array}$$

Answer: 29.3760

Using method 4:

$C = D^2 \times L \times 0.0034$

When measurements are in inches:

1. $\begin{array}{r} 12 \\ \underline{\times 12} \\ 144 \end{array}$ 2. $\begin{array}{r} 144 \\ \underline{\times 60} \\ 8640 \end{array}$ 3. $\begin{array}{r} 8640 \\ \underline{\times .0034} \\ 34560 \\ \underline{25920} \end{array}$

Answer: 29.3760

Only three steps are necessary using this method; also methods 3 and 4 arrive at the same exact answer when carried to four decimal points.

Problem

A well casing is 4 inches in diameter, 120 feet deep, and has 10 feet of water standing in the casing. How many gallons of water are in the casing?

Using method 3:

1. $\begin{array}{r} 4 \\ \underline{\times 4} \\ 16 \end{array}$ 2. $\begin{array}{r} 16 \\ \underline{\times 10} \\ 160 \end{array}$ 3. $\begin{array}{r} 160 \\ \underline{\times .0408} \\ 1280 \\ \underline{6400} \\ 6.5280 \end{array}$

Using method 4:

1. $\begin{array}{r} 4 \\ \underline{\times 4} \\ 16 \end{array}$ 2. $\begin{array}{r} 16 \\ \underline{\times 120} \\ 320 \end{array}$ 3. $\begin{array}{r} 1920 \\ \underline{\times .0034} \\ 7680 \\ \underline{5760} \\ 6.5280 \end{array}$

Using either feet times inches or inches times inches, the answers are the same.

Fractional Equivalents of Decimals

Decimal	Fraction	Decimal	Fraction
.015625	1/64	.4375	7/16
.03125	1/32	.453125	29/64
.046875	3/64	.46875	15/32
.05	1/20	.484375	31/64
.0625	1/16	.5	1/2
.07693	1/13	.515625	33/64
.078125	5/64	.53125	17/32
.0833	1/12	.546875	35/64
.0909	1/11	.5625	9/16
.093753	3/32	.578125	37/64
.10	1/10	.59375	9/32
.109375	7/64	.609375	39/64
.111	1/9	.625	5/8
.125	1/8	.640625	41/54
.14062	59/64	.65625	21/32
.1429	1/7	.671875	43/64
.15625	5/32	.6875	11/16
.1667	1/6	.703125	45/64
.171875	11/64	.71875	23/32
.1875	3/16	.734375	47/64
.2	1/5	.75	3/4
.203125	13/64	.765625	49/64
.21875	7/32	.78125	25/32
.234375	15/64	.796875	51/64
.25	1/4	.8125	13/16
.265625	17/64	.828125	53/64
.28125	9/32	.84375	27/32
.296875	19/64	.859375	55/64
.333	1/3	.90625	29/32
.34375	11/32	.921875	59/64
.359375	23/54	.9375	15/16
.375	3/8	.953125	61/64
.390625	25/64	.96875	31/32
.40625	13/32	.984375	63/64
.421875	27/64	1	1

Building Codes

Building codes regulate everything about the buildings in which we live and work. Different organizations develop codes and regulations that are then adopted by cities, states, or counties. Contact the building department in your city, state, or county to find out which code is in effect in your area.

Building codes are not cast in stone. They are *minimum* standards. A state may adopt a code or regulations, and a city within that state may change a few words in a section of definitions, specifications, or requirements and in doing so may tighten the requirements. New versions that have been adopted by a city, county, or state may contain slightly different wording and meaning from previous versions. For this reason I caution you throughout this book to not take anything for granted. Always check with the building official in the area where the work is being done to ensure that the work contemplated will comply with the building codes and regulations.

I have acquired quite a collection of various plumbing code books over the years; they grow in size and content with each new version. Building codes and regulations must be observed and enforced both in new construction and remodeling of existing buildings.

For information about building codes adopted and enforced in various areas of the United States and Canada as well as in other parts of the world, the following list of organizations with addresses, telephone numbers, and/or web sites and fax numbers is provided:

American National Standards Institute (ANSI)
Headquarters: 1819 L Street, NW, Washington, DC 20036
Phone: (202) 293-8020
Fax: (202) 293-9287
New York office 11 W. 42nd Street
New York, NY 10031

Call for list of numerous codes and regulations.

International Association of Plumbing and Mechanical Officials (IAPMO)
20001 E. Walnut Dr., South
Walnut, CA 91789-2825
Phone: (909) 595-84491
General e-mail: iapmo@iapmo.org.

The following Codes and Regulations are available from IAPMO: Uniform Plumbing Code© (UPC), Uniform Mechanical Code© (UMC), Uniform Solar Energy Code© (USEC), Uniform Swimming Pool, SPA, and Hot Tub Code© (USPC), ANSI A-40 Plumbing Code, Uniform Fire Code©. IAPMO codes have been adopted in more than 35 states and foreign countries.

Southern Building Code Congress International Inc.
900 Montclair Road, Birmingham, AL 35213-1206
Phone: (205) 591-1853
Fax: (205) 591-0775

The Southern Building Code Congress International, Inc. (SBCCI) provides technical, educational, and administrative support to governmental departments and agencies engaged in building codes administration and enforcement. SBCCI codes and regulations have been adopted in many states. For information on obtaining copies of Standard Building Codes, including all phases of construction, call the number listed.

American Gas Association (AGA)
400 N. Capitol St. NW
Washington, DC 20001
Phone: (202) 824-7000
Website: http://www.aga.org

AGA has available to members and non-members at competitive prices the soft cover *National Fuel Gas Code Handbook* and the *National Fuel Gas Code* set, soft cover *Code* plus the *Handbook*.

American Society of Mechanical Engineers
Three Park Avenue
New York, NY
10016-5990
Phones: (800) 843 ASME (United States and Canada), (978) 882-1167 (outside North America)
e-mail: infocentral@asme.org

National Fire Protection Association, Inc.
1 Batterymarch Park
P.O. Box 9101
Quincy, MA 02269-9101
Phone: (617) 770-3000
Fax: (617) 770-0700

Call for information on codes.

BOCA International
4051 N. Flossmor Rd.
Country Club Hills, IL 60478
Phone: (708) 799-2300

Underwriters Laboratories, Inc.
330 Pfingsten Road
Northbrook, IL 60062-2096
e-mail: northbrook@us.ul.com

Offices in Asia, South America, Australia, Brazil

Radiator Specialty Company provides high quality replacement parts and accessories that are either manufactured or procured by us. Original manufacturer's part number or name, and/or brand and model names, are used for the purpose of identification only, and are not a warranty or representation, either express or implied, that items offered or sold are genuine products of such original manufacturers.

REPLACEMENT FAUCET STEMS

To identify your stem, simply place it on the illustrations until a close match is found. This should be your stem number. The text accompanying each illustration includes original manufacturer, OEM part number and if hot or cold sides are different. If number includes "C" (FS1-9C), this is cold side. If number includes "H" (FS1-9H), this is hot side. If number includes "HC" (FS1-13HC), stem fits either side.

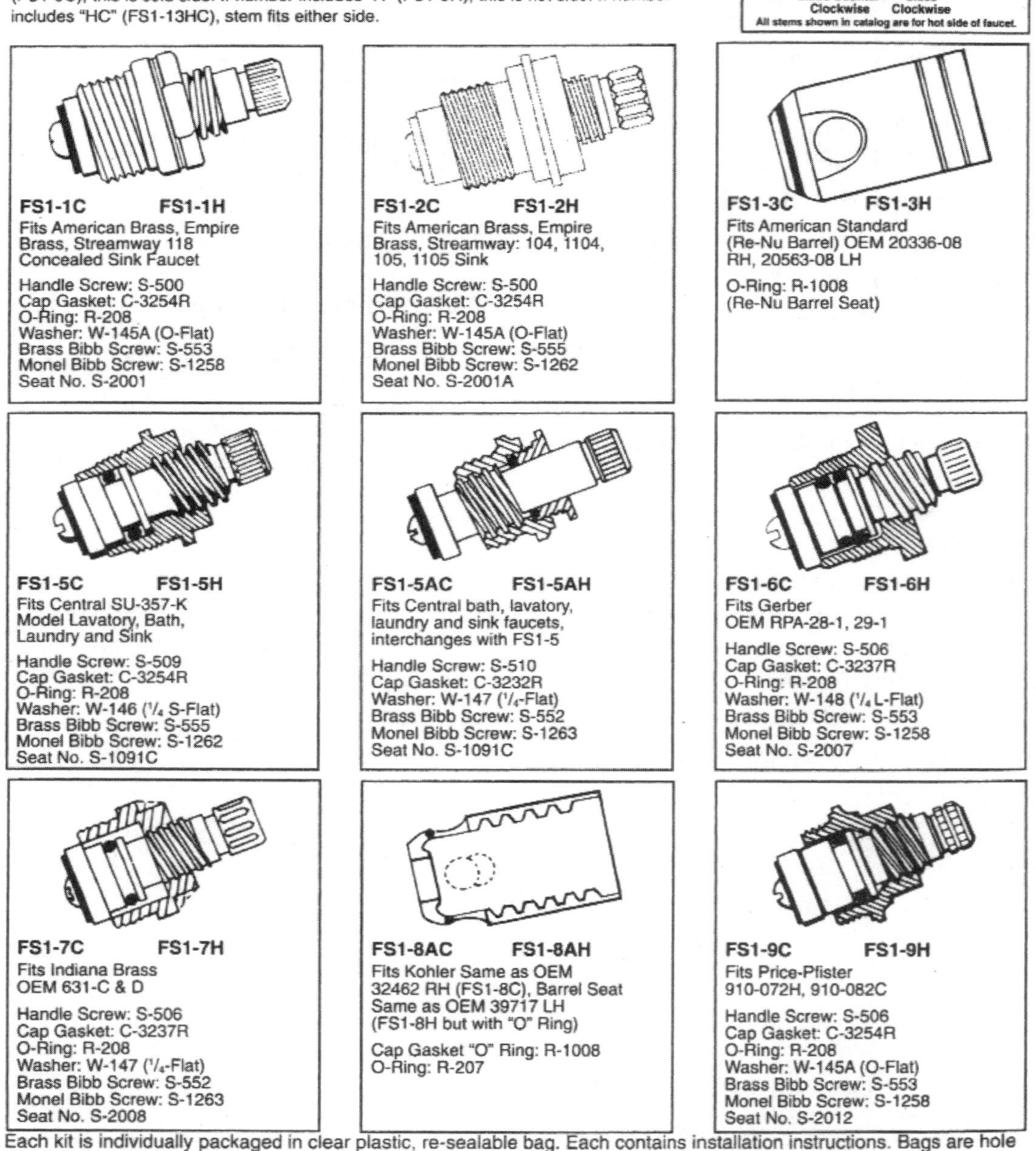

FS1-1C **FS1-1H**
Fits American Brass, Empire Brass, Streamway 118 Concealed Sink Faucet

Handle Screw: S-500
Cap Gasket: C-3254R
O-Ring: R-208
Washer: W-145A (O-Flat)
Brass Bibb Screw: S-553
Monel Bibb Screw: S-1258
Seat No. S-2001

FS1-2C **FS1-2H**
Fits American Brass, Empire Brass, Streamway: 104, 1104, 105, 1105 Sink

Handle Screw: S-500
Cap Gasket: C-3254R
O-Ring: R-208
Washer: W-145A (O-Flat)
Brass Bibb Screw: S-555
Monel Bibb Screw: S-1262
Seat No. S-2001A

FS1-3C **FS1-3H**
Fits American Standard (Re-Nu Barrel) OEM 20336-08 RH, 20563-08 LH

O-Ring: R-1008
(Re-Nu Barrel Seat)

FS1-5C **FS1-5H**
Fits Central SU-357-K Model Lavatory, Bath, Laundry and Sink

Handle Screw: S-509
Cap Gasket: C-3254R
O-Ring: R-208
Washer: W-146 (1/4 S-Flat)
Brass Bibb Screw: S-555
Monel Bibb Screw: S-1262
Seat No. S-1091C

FS1-5AC **FS1-5AH**
Fits Central bath, lavatory, laundry and sink faucets, interchanges with FS1-5

Handle Screw: S-510
Cap Gasket: C-3232R
Washer: W-147 (1/4-Flat)
Brass Bibb Screw: S-552
Monel Bibb Screw: S-1263
Seat No. S-1091C

FS1-6C **FS1-6H**
Fits Gerber OEM RPA-28-1, 29-1

Handle Screw: S-506
Cap Gasket: C-3237R
O-Ring: R-208
Washer: W-148 (1/4 L-Flat)
Brass Bibb Screw: S-553
Monel Bibb Screw: S-1258
Seat No. S-2007

FS1-7C **FS1-7H**
Fits Indiana Brass OEM 631-C & D

Handle Screw: S-506
Cap Gasket: C-3237R
O-Ring: R-208
Washer: W-147 (1/4-Flat)
Brass Bibb Screw: S-552
Monel Bibb Screw: S-1263
Seat No. S-2008

FS1-8AC **FS1-8AH**
Fits Kohler Same as OEM 32462 RH (FS1-8C), Barrel Seat Same as OEM 39717 LH (FS1-8H but with "O" Ring)

Cap Gasket "O" Ring: R-1008
O-Ring: R-207

FS1-9C **FS1-9H**
Fits Price-Pfister 910-072H, 910-082C

Handle Screw: S-506
Cap Gasket: C-3254R
O-Ring: R-208
Washer: W-145A (O-Flat)
Brass Bibb Screw: S-553
Monel Bibb Screw: S-1258
Seat No. S-2012

Each kit is individually packaged in clear plastic, re-sealable bag. Each contains installation instructions. Bags are hole punched for peg display.

Courtesy Radiator Specialty Co.

Radiator Specialty Company provides high quality replacement parts and accessories that are either manufactured or procured by us. Original manufacturer's part number or name, and/or brand and model names, are used for the purpose of identification only, and are not a warranty or representation, either express or implied, that items offered or sold are genuine products of such original manufacturers.

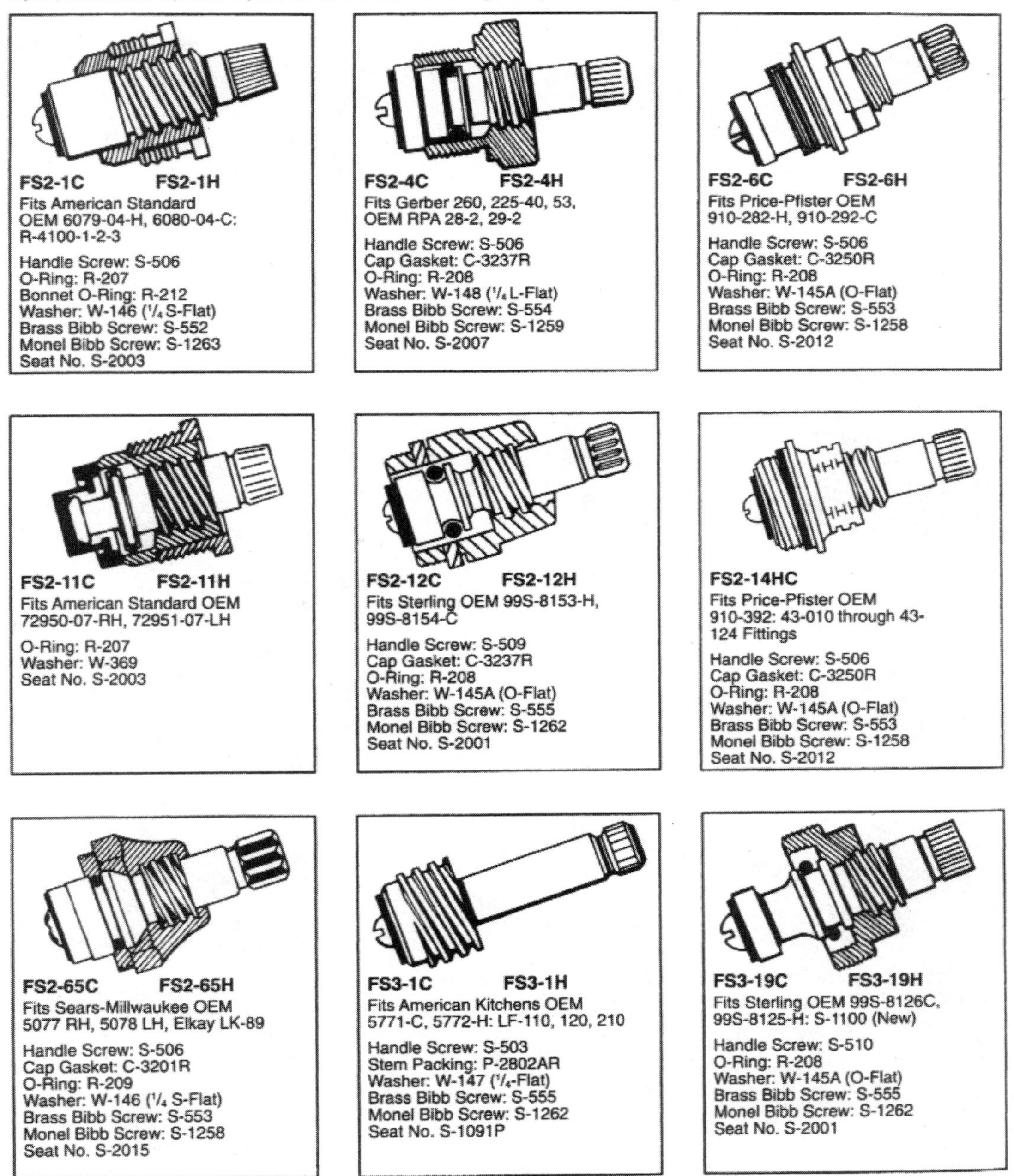

FS2-1C FS2-1H
Fits American Standard OEM 6079-04-H, 6080-04-C: R-4100-1-2-3
Handle Screw: S-506
O-Ring: R-207
Bonnet O-Ring: R-212
Washer: W-146 (1/4 S-Flat)
Brass Bibb Screw: S-552
Monel Bibb Screw: S-1263
Seat No. S-2003

FS2-4C FS2-4H
Fits Gerber 260, 225-40, 53, OEM RPA 28-2, 29-2
Handle Screw: S-506
Cap Gasket: C-3237R
O-Ring: R-208
Washer: W-148 (1/4 L-Flat)
Brass Bibb Screw: S-554
Monel Bibb Screw: S-1259
Seat No. S-2007

FS2-6C FS2-6H
Fits Price-Pfister OEM 910-282-H, 910-292-C
Handle Screw: S-506
Cap Gasket: C-3250R
O-Ring: R-208
Washer: W-145A (O-Flat)
Brass Bibb Screw: S-553
Monel Bibb Screw: S-1258
Seat No. S-2012

FS2-11C FS2-11H
Fits American Standard OEM 72950-07-RH, 72951-07-LH
O-Ring: R-207
Washer: W-369
Seat No. S-2003

FS2-12C FS2-12H
Fits Sterling OEM 99S-8153-H, 99S-8154-C
Handle Screw: S-509
Cap Gasket: C-3237R
O-Ring: R-208
Washer: W-145A (O-Flat)
Brass Bibb Screw: S-555
Monel Bibb Screw: S-1262
Seat No. S-2001

FS2-14HC
Fits Price-Pfister OEM 910-392: 43-010 through 43-124 Fittings
Handle Screw: S-506
Cap Gasket: C-3250R
O-Ring: R-208
Washer: W-145A (O-Flat)
Brass Bibb Screw: S-553
Monel Bibb Screw: S-1258
Seat No. S-2012

FS2-65C FS2-65H
Fits Sears-Millwaukee OEM 5077 RH, 5078 LH, Elkay LK-89
Handle Screw: S-506
Cap Gasket: C-3201R
O-Ring: R-209
Washer: W-146 (1/4 S-Flat)
Brass Bibb Screw: S-553
Monel Bibb Screw: S-1258
Seat No. S-2015

FS3-1C FS3-1H
Fits American Kitchens OEM 5771-C, 5772-H: LF-110, 120, 210
Handle Screw: S-503
Stem Packing: P-2802AR
Washer: W-147 (1/4-Flat)
Brass Bibb Screw: S-555
Monel Bibb Screw: S-1262
Seat No. S-1091P

FS3-19C FS3-19H
Fits Sterling OEM 99S-8126C, 99S-8125-H: S-1100 (New)
Handle Screw: S-510
O-Ring: R-208
Washer: W-145A (O-Flat)
Brass Bibb Screw: S-555
Monel Bibb Screw: S-1262
Seat No. S-2001

Each kit is individually packaged in clear plastic, re-sealable bag. Each contains installation instructions. Bags are hole punched for peg display.

Courtesy Radiator Specialty Co.

FS4-25C FS4-25H
Fits Eljer OEM 4788: No. 3 Unit

Handle Screw: S-506
Cap Gasket: C-3265R
Stem Packing: P-2833R
Washer: W-147 (1/4-Flat)
Brass Bibb Screw: S-555
Monel Bibb Screw: S-1262
Seat No. S-1091J

FS4-27C FS4-27H
Fits American Standard OEM 64703-07-RH, 50668-07-LH: Colony Trim

Handle Screw: S-506
Stem Packing: P-2823R
Washer: W-146 (1/4 S-Flat)
Brass Bibb Screw: S-555
Monel Bibb Screw: S-1262
Seat No. S-1091R

FS4-28C FS4-28H
Fits Sears, Elkay, Universal-Rundle OEM P-1077, 32008-L

Handle Screw: S-506
Cap Thread "O" Ring: R-212
Stem "O" Ring: R-208
Bibb-Seat "O" Ring: R-207
Washer: W-146 (1/4 S-Flat)
Brass Bibb Screw: S-553
Monel Bibb Screw: S-1258

FS4-29C FS4-29H
Fits Price-Pfister OEM 910-492

Handle Screw: S-507
Cap Gasket: C-3265R
O-Ring: R-208
Washer: W-145 (OO-Flat)
Brass Bibb Screw: S-553
Monel Bibb Screw: S-1258
Seat No. S-2012

FS4-30HC
Fits Price-Pfister OEM 910-681

Handle Screw: S-507
Cap Gasket: C-3250R
O-Ring: R-208

FS4-61HC
Fits Eljer OEM 5182

Handle Screw: S-501
Cap Gasket: C-3254R
Stem Packing: P-2824R
Washer: W-147 (1/4-Flat)
Brass Bibb Screw: S-555
Monel Bibb Screw: S-1262
Seat No. S-2032

FS4-62C FS4-62H
Fits Eljer OEM 4650

Handle Screw: S-506
Cap Gasket: C-3265R
Stem Packing: P-2833R
Washer: W-146 (1/4 S-Flat)
Brass Bibb Screw: S-555
Monel Bibb Screw: S-1262
Seat No. S-2032

FS4-63C FS4-63H
Fits Kohler OEM 20655-H, 20656-C: Valve Unit, Aquaric

Handle Screw: S-500
Cap Thread "O" Ring: R-994
Stem "O" Ring: R-209
Washer: W-145A (O-Flat)
Brass Bibb Screw: S-550
Seat No. S-2022

FS4-72C FS4-72H
Fits Eljer OEM 2807: 9575-R, 9576-R

Handle Screw: S-506
Cap Gasket: C-3263R
Stem Packing: P-2811R
Washer: W-147 (1/4-Flat)
Brass Bibb Screw: S-552
Monel Bibb Screw: S-1263
Seat No. S-2032

Each kit is individually packaged in clear plastic, re-sealable bag. Each contains installation instructions. Bags are hole punched for peg display.

Courtesy Radiator Specialty Co.

Radiator Specialty Company provides high quality replacement parts and accessories that are either manufactured or procured by us. Original manufacturer's part number or name, and/or brand and model names, are used for the purpose of identification only, and are not a warranty or representation, either express or implied, that items offered or sold are genuine products of such original manufacturers.

FS6-5C FS6-5H
Fits Crane OEM F13167-RH, F13168-LH Hospital Stem

Cap Gasket: C-3207AR
O-Ring: R-1004
Washer: W-356

FS6-63C FS6-63H
Fits Savoy Brass OEM A-17-H, A-17-C: 18 Serration Broach, New Style; also Richmond

Handle Screw: S-506
Cap Gasket: C-3254R
Stem Packing: P-2822R
Washer: W-145A (O-Flat)
Brass Bibb Screw: S-553
Monel Bibb Screw: S-1258
Seat No. 2026

FS6-65C FS6-65H
Fits Eljer OEM 2733: E-9340-R-41R Conc. Lavatory Fittings

Cap Gasket: C-3212R
Stem Packing: P-2824R
Washer: W-146 (1/4 S-Flat)
Brass Bibb Screw: S-555
Monel Bibb Screw: S-1262
Seat No. 2032

FS7-1HC
Fits American Standard OEM 19376-2: F-105, F-115 Lavatory, B-787 Lavatory, B-912S Sink

Handle Screw: S-501
Stem Packing: P-2800R
Washer: W-146 (1/4 S-Flat)
Brass Bibb Screw: S-553
Monel Bibb Screw: S-1258
Seat No. FS1-3H

FS7-10C FS7-10H
Fits Kohler OEM 31584-C, 31585-H: Hapton, Taughton, Marston, Gram., Strand, K-8100-15-32-22

Handle Screw: S-500
Cap Gasket: C-3223R
O-Ring: R-1004
Washer: W-145A (O-Flat)
Brass Bibb Screw: S-555
Monel Bibb Screw: S-1262
Seat No. S-1090

FS7-17C FS7-17H
Fits Union Brass OEM 1840-A-RH, 1840-A-LH: 30-2-3-4-5

Handle Screw: S-507
Cap Gasket: C-3232R
O-Ring: R-208
Washer: W-146 (1/4 S-Flat)
Brass Bibb Screw: S-555
Monel Bibb Screw: S-1262
Seat No. S-2028

Each kit is individually packaged in clear plastic, re-sealable bag. Each contains installation instructions. Bags are hole punched for peg display.

Courtesy Radiator Specialty Co.

Radiator Specialty Company provides high quality replacement parts and accessories that are either manufactured or procured by us. Original manufacturer's part number or name, and/or brand and model names, are used for the purpose of identification only, and are not a warranty or representation, either express or implied, that items offered or sold are genuine products of such original manufacturers.

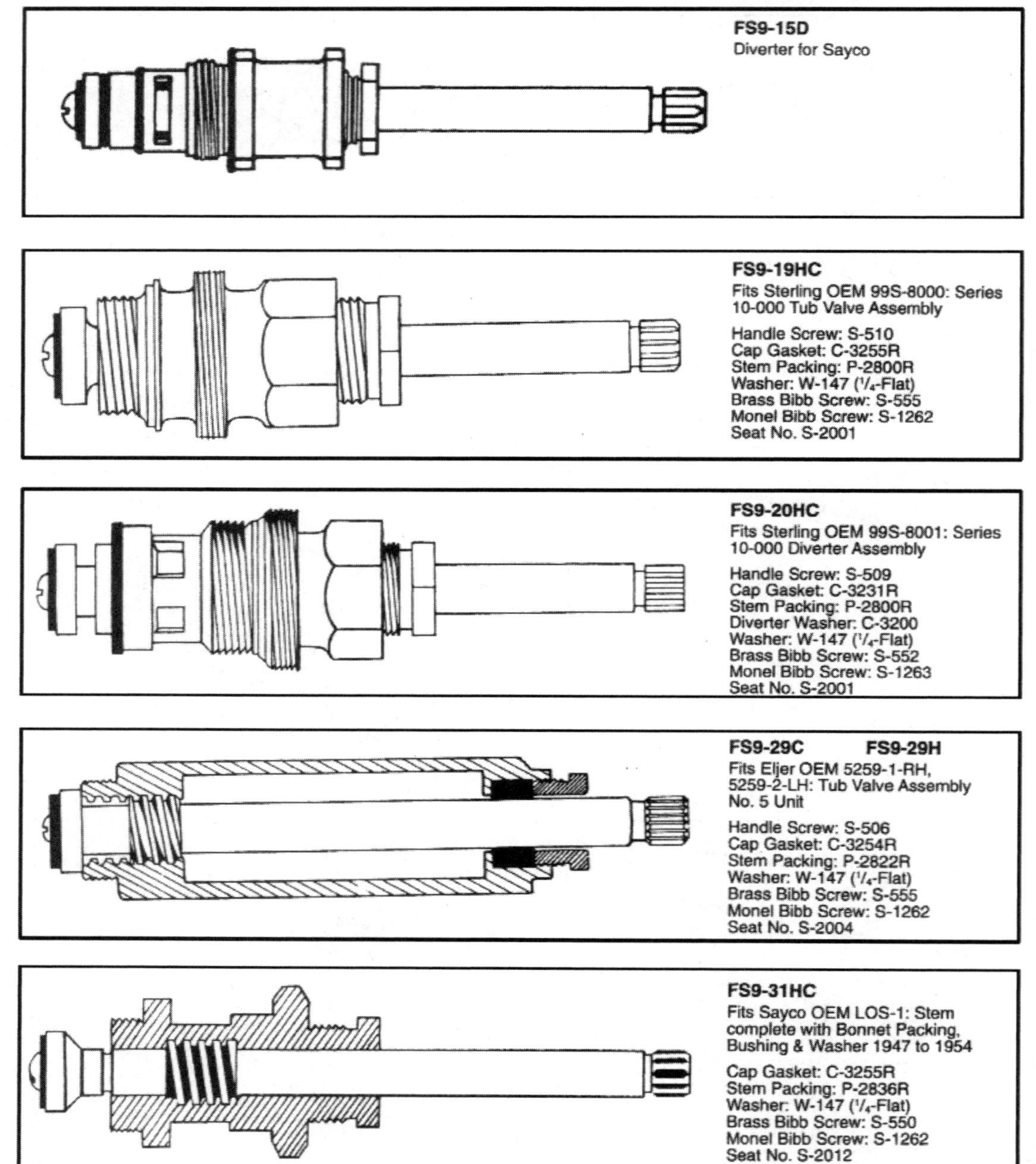

FS9-15D
Diverter for Sayco

FS9-19HC
Fits Sterling OEM 99S-8000: Series 10-000 Tub Valve Assembly

Handle Screw: S-510
Cap Gasket: C-3255R
Stem Packing: P-2800R
Washer: W-147 (1/4-Flat)
Brass Bibb Screw: S-555
Monel Bibb Screw: S-1262
Seat No. S-2001

FS9-20HC
Fits Sterling OEM 99S-8001: Series 10-000 Diverter Assembly

Handle Screw: S-509
Cap Gasket: C-3231R
Stem Packing: P-2800R
Diverter Washer: C-3200
Washer: W-147 (1/4-Flat)
Brass Bibb Screw: S-552
Monel Bibb Screw: S-1263
Seat No. S-2001

FS9-29C **FS9-29H**
Fits Eljer OEM 5259-1-RH, 5259-2-LH: Tub Valve Assembly No. 5 Unit

Handle Screw: S-506
Cap Gasket: C-3254R
Stem Packing: P-2822R
Washer: W-147 (1/4-Flat)
Brass Bibb Screw: S-555
Monel Bibb Screw: S-1262
Seat No. S-2004

FS9-31HC
Fits Sayco OEM LOS-1: Stem complete with Bonnet Packing, Bushing & Washer 1947 to 1954

Cap Gasket: C-3255R
Stem Packing: P-2836R
Washer: W-147 (1/4-Flat)
Brass Bibb Screw: S-550
Monel Bibb Screw: S-1262
Seat No. S-2012

Each kit is individually packaged in clear plastic, re-sealable bag. Each contains installation instructions. Bags are hole punched for peg display.

Courtesy Radiator Specialty Co.

Radiator Specialty Company provides high quality replacement parts and accessories that are either manufactured or procured by us. Original manufacturer's part number or name, and/or brand and model names, are used for the purpose of identification only, and are not a warranty or representation, either express or implied, that items offered or sold are genuine products of such original manufacturers.

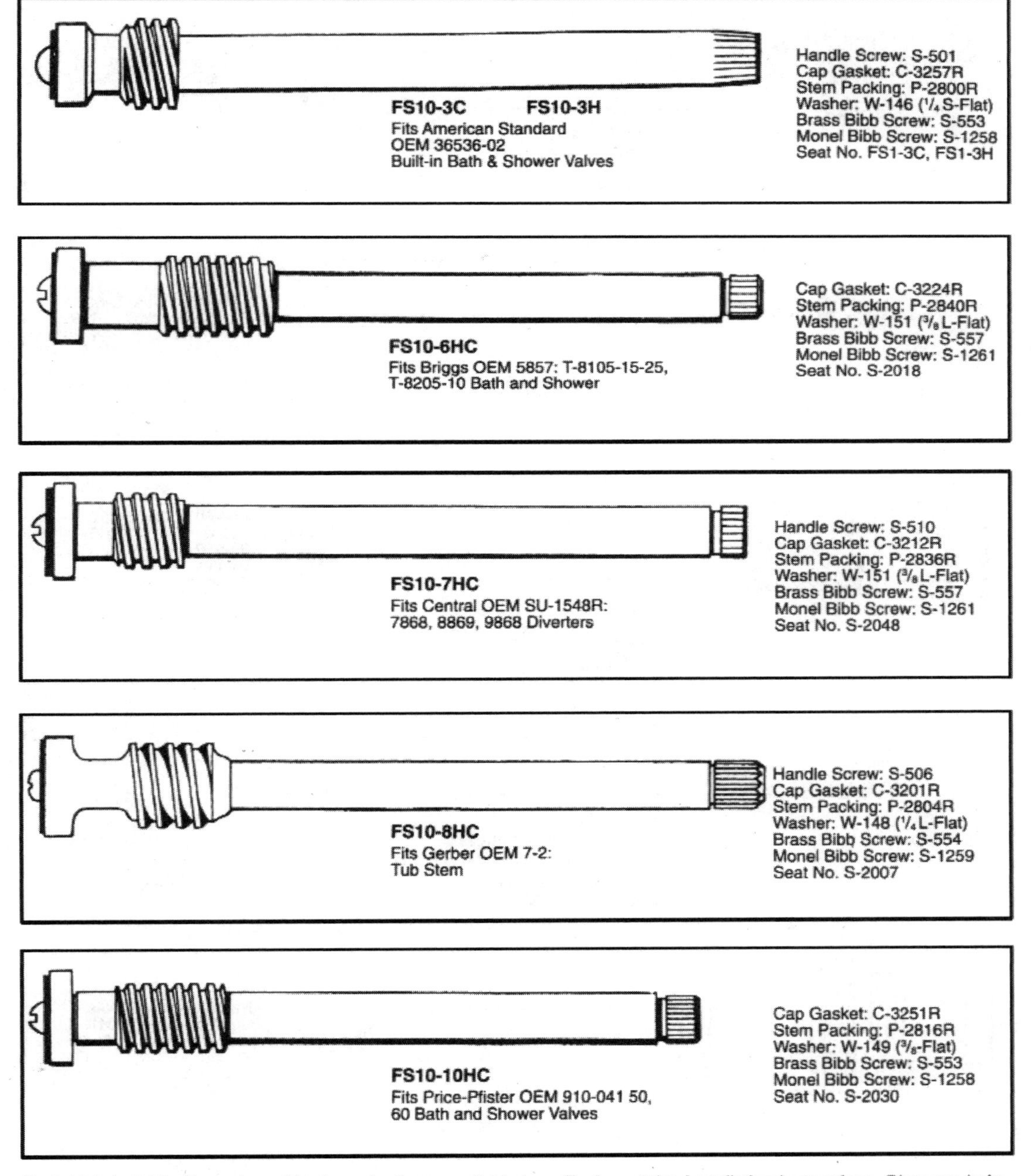

Each kit is individually packaged in clear plastic, re-sealable bag. Each contains installation instructions. Bags are hole punched for peg display.

Courtesy Radiator Specialty Co.

Radiator Specialty Company provides high quality replacement parts and accessories that are either manufactured or procured by us. Original manufacturer's part number or name, and/or brand and model names, are used for the purpose of identification only, and are not a warranty or representation, either express or implied, that items offered or sold are genuine products of such original manufacturers.

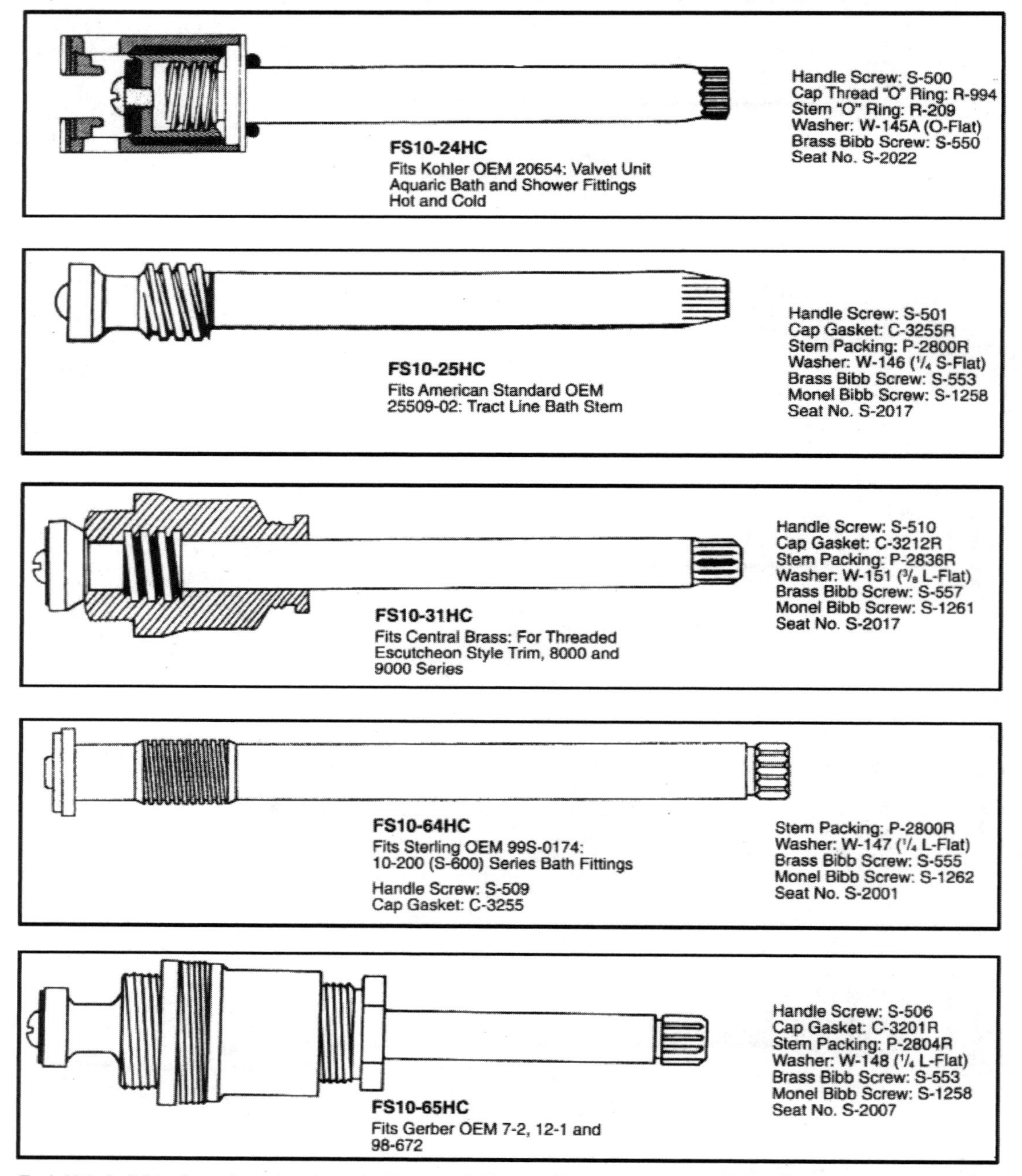

Each kit is individually packaged in clear plastic, re-sealable bag. Each contains installation instructions. Bags are hole punched for peg display.

Courtesy Radiator Specialty Co.

Radiator Specialty Company provides high quality replacement parts and accessories that are either manufactured or procured by us. Original manufacturer's part number or name, and/or brand and model names, are used for the purpose of identification only, and are not a warranty or representation, either express or implied, that items offered or sold are genuine products of such original manufacturers.

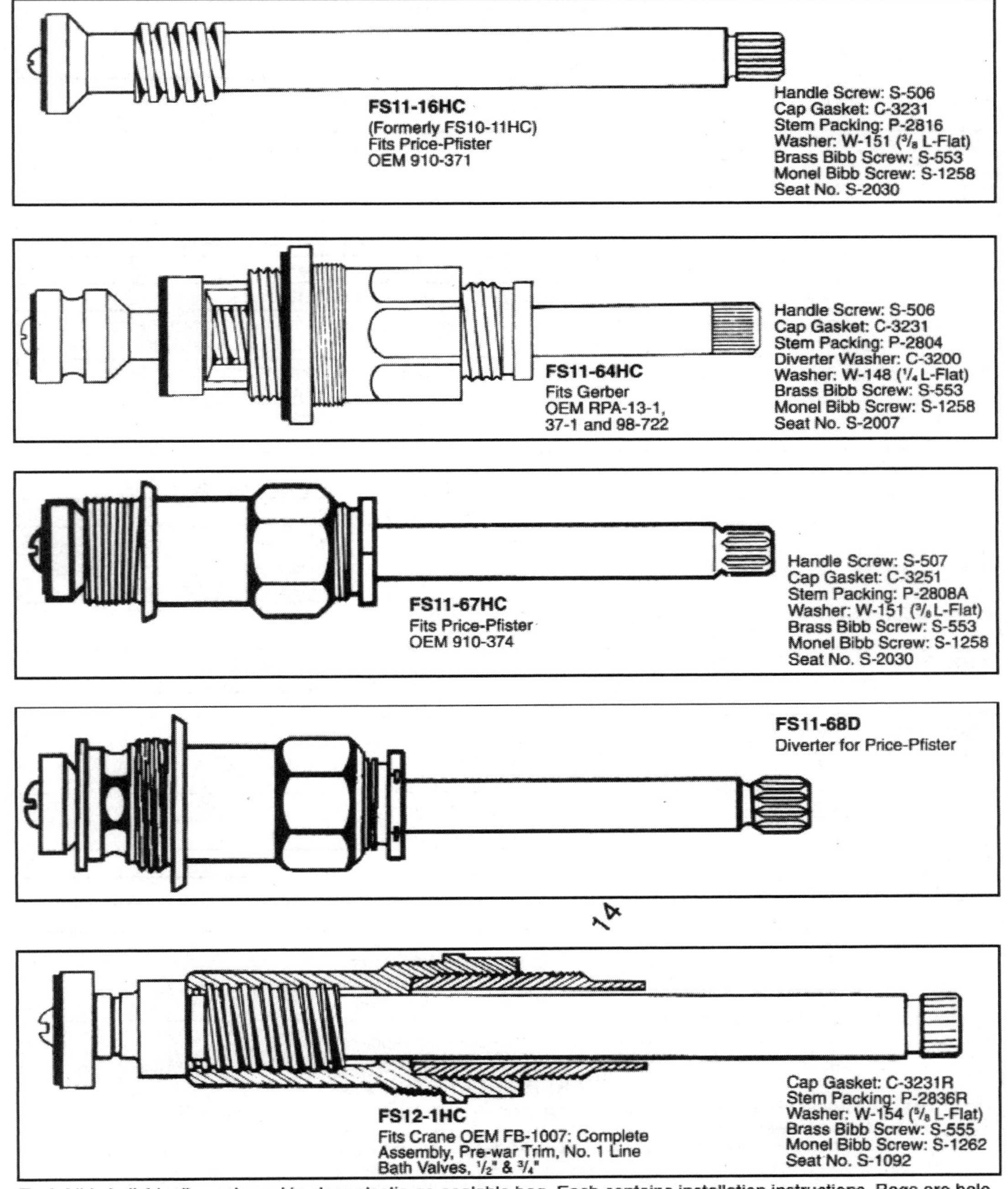

Each kit is individually packaged in clear plastic, re-sealable bag. Each contains installation instructions. Bags are hole punched for peg display

Courtesy Radiator Specialty Co.

Radiator Specialty Company provides high quality replacement parts and accessories that are either manufactured or procured by us. Original manufacturer's part number or name, and/or brand and model names, are used for the purpose of identification only, and are not a warranty or representation, either express or implied, that items offered or sold are genuine products of such original manufacturers.

FAUCET REPAIR KITS

Part No.	Faucet Replacement Parts For:	O.E.M. No.	Description	Contains
K-193	Chicago Faucet	889		1 ea: P-2228 2 ea: P-2815, C-3206A, L-2619, L-2621, W-405
K-240	Crane	All Dial-Ese		2 ea: C-3207A, C-3226, R-1004, W-356
K-258	For Delta*, Delex* Faucet		Brass Ball, Lever Handle	1 ea: Brass Ball
K-260	For Delta*, Delex* Faucet		Plastic Ball, Lever Handle	1 ea: Plastic Ball Note: K-258 & K-260 fit most lever type units except bath valve mfd. before 1968
K-262	For Delta*, Delex* Faucet		Brass Ball, Knob Handle	1 ea: Brass Ball
K-268-S	For Delta*, Delex* Faucet		Seats-Springs (Old Style)	2 ea: C-3210, L-2611
K-269-S	For Delta*, Delex* Faucet		Complete Kit (Old Style)	1 ea: C-3238A, P-2232, T-2021, T-2022 2 ea: C-3210, L-2611, R-991, R-223
K-270-S	For Delta*, Delex* Faucet		Complete Kit for Knob Handle (Old Style)	1 ea: C-3238A, P-2232A, T-2021, T-2022 2 ea: C-3210, L-2611, R-223, R-991
K-271-S	For Delta*, Delex* Faucet		Seats-Springs (New Style, Since 1969)	2 ea: C-3271, L-2640
K-272-S	For Delta*, Delex* Faucet		Complete Kit for Lever Handle (New Style, Since 1969)	1 ea: C-3238A, P-2232, T-2021, T-2022 2 ea: C-3271, L-2640, R-991, R-223
K-273	For Delta*, Delex* Faucet		Cartridge	1 ea: Cartridge
K-420-S	Gerber		Faucet Repair Kit w/Seats	2 ea: C3214, R208, R214, S-1259, W-278, S-2007
K-792-S	Schaible			1 ea: L-2602, R-992, R-1006 2 ea: L-2617, L-2636, C-3228, S-1091S1
K-1200	Moen (Genuine)	All Moen models Made since 1959		1 ea: K-1200 Brass Cartridge
K-1225	Moen-Chateau (Genuine)	All Moen and Chateau models made since 1959		1 ea: K-1225 Plastic Cartridge Note: K-1225 is a replacement cartridge for K-1200 & K-1201
C-3210	For Delta*, Delex* Faucet		Seat (Old Style)	
C-3238A	For Delta*, Delex* Faucet		Cap Gasket - use with P-2232 & P-2232A	
C-3271	For Delta*, Delex* Faucet		Seat (New Style) 1969 & after	
L-2640	For Delta*, Delex* Faucet		Spring (New Style) 1969 & after	
P-2232	For Delta*, Delex* Faucet		Lever Type	
P-2232A	For Delta*, Delex* Faucet		Knob Type	
R-223	For Delta*, Delex* Faucet		O-Ring	
R-991	For Delta*, Delex* Faucet		O-Ring	

FOR DELTA*, DELEX* OR PEERLESS* FAUCETS
THESE PARTS ARE NOT MADE BY DELTA

RADIATOR SPECIALTY COMPANY

***CAUTION: THIS PRODUCT IS NOT MANUFACTURED OR AUTHORIZED BY THE MAKER OF DELTA (OR PEERLESS OR DELEX) FAUCETS, WHICH HAS NO RESPONSIBILITY FOR THE QUALITY OR SUITABILITY OF THESE REPAIR PARTS. USE OF THESE PARTS IN A DELTA (OR DELEX OR PEERLESS) FAUCET MAY VOID THE DELTA (OR DELEX OR PEERLESS) WARRANTY.**

Courtesy Radiator Specialty Co.

Radiator Specialty Company provides high quality replacement parts and accessories that are either manufactured or procured by us. Original manufacturer's part number or name, and/or brand and model names, are used for the purpose of identification only, and are not a warranty or representation, either express or implied, that items offered or sold are genuine products of such original manufacturers.

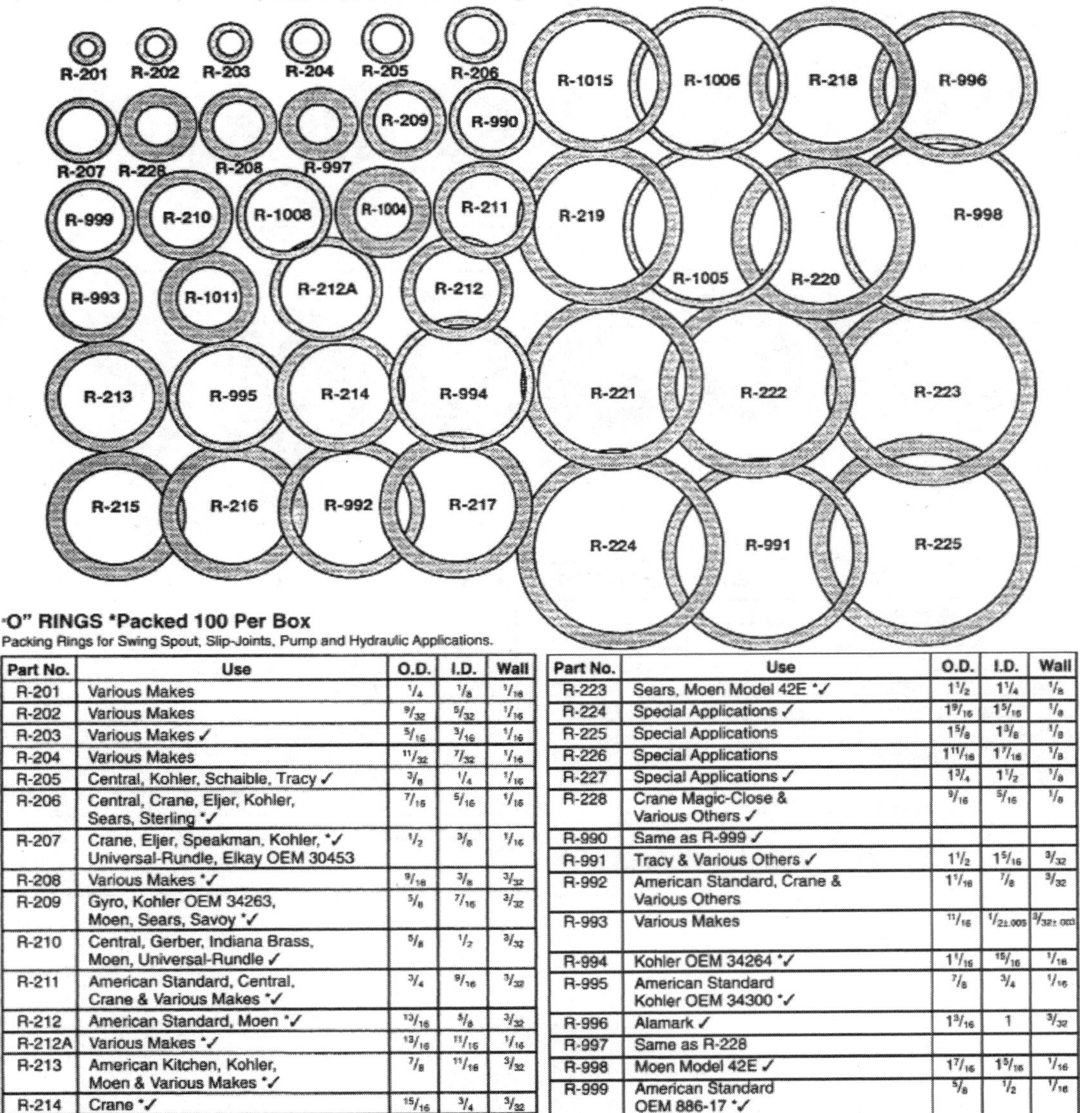

"O" RINGS *Packed 100 Per Box

Packing Rings for Swing Spout, Slip-Joints, Pump and Hydraulic Applications.

Part No.	Use	O.D.	I.D.	Wall
R-201	Various Makes	1/4	1/8	1/16
R-202	Various Makes	9/32	5/32	1/16
R-203	Various Makes ✓	5/16	3/16	1/16
R-204	Various Makes	11/32	7/32	1/16
R-205	Central, Kohler, Schaible, Tracy ✓	3/8	1/4	1/16
R-206	Central, Crane, Eljer, Kohler, Sears, Sterling *✓	7/16	5/16	1/16
R-207	Crane, Eljer, Speakman, Kohler, *✓ Universal-Rundle, Elkay OEM 30453	1/2	3/8	1/16
R-208	Various Makes *✓	9/16	3/8	3/32
R-209	Gyro, Kohler OEM 34263, Moen, Sears, Savoy *✓	5/8	7/16	3/32
R-210	Central, Gerber, Indiana Brass, Moen, Universal-Rundle ✓	5/8	1/2	3/32
R-211	American Standard, Central, Crane & Various Makes *✓	3/4	9/16	3/32
R-212	American Standard, Moen *✓	13/16	5/8	3/32
R-212A	Various Makes *✓	13/16	11/16	1/16
R-213	American Kitchen, Kohler, Moen & Various Makes *✓	7/8	11/16	3/32
R-214	Crane *✓	15/16	3/4	3/32
R-215	Crane ✓	1	3/4	1/8
R-216	Eljer, Gyro, Moen *✓	1 1/16	13/16	1/8
R-217	American Standard OEM 507-37, Moen 137, Repcal ✓	1 1/8	7/8	1/8
R-218	Speakman	1 3/16	15/16	1/8
R-219	Various Makes	1 1/4	1	1/8
R-220	Various Makes ✓	1 5/16	1 1/16	1/8
R-221	Moen Model 52 & Various Makes	1 3/8	1 1/8	1/8
R-222	Various Makes	1 7/16	1 3/16	1/8
R-223	Sears, Moen Model 42E *✓	1 1/2	1 1/4	1/8
R-224	Special Applications ✓	1 9/16	1 5/16	1/8
R-225	Special Applications	1 5/8	1 3/8	1/8
R-226	Special Applications	1 11/16	1 7/16	1/8
R-227	Special Applications ✓	1 3/4	1 1/2	1/8
R-228	Crane Magic-Close & Various Others ✓	9/16	5/16	1/8
R-990	Same as R-999 ✓			
R-991	Tracy & Various Others ✓	1 1/2	1 5/16	3/32
R-992	American Standard, Crane & Various Others	1 1/16	7/8	3/32
R-993	Various Makes	11/16	1/2 ±.005	3/32 ±.003
R-994	Kohler OEM 34264 *✓	1 1/16	15/16	1/16
R-995	American Standard Kohler OEM 34300 *✓	7/8	3/4	1/16
R-996	Alamark ✓	1 3/16	1	3/32
R-997	Same as R-228			
R-998	Moen Model 42E ✓	1 7/16	1 5/16	1/16
R-999	American Standard OEM 886-17 *✓	5/8	1/2	1/16
R-1004	Crane OEM F12376 ✓	45/64	27/64	9/64
R-1005	Various Makes	1 1/4	1 1/8	1/16
R-1006	American Std. OEM 12035, Schaible, Sears, Youngstown	1 3/16	1 1/16	1/16
R-1008	Harcraft *✓	11/16	9/16	1/16
R-1011	Barnes ✓	23/32	13/32	1/8
R-1015	Eljer OEM 5285, Lusterline	1 1/8	15/16	3/32
R-1016	Am. Standard Aquarian Faucet	2 3/16	2	3/32
R-292	100 Assortment, 12 Various Sizes			

Courtesy Radiator Specialty Co.

Radiator Specialty Company provides high quality replacement parts and accessories that are either manufactured or procured by us. Original manufacturer's part number or name, and/or brand and model names, are used for the purpose of identification only, and are not a warranty or representation, either express or implied, that items offered or sold are genuine products of such original manufacturers.

GOLDEN STATE® FAUCET WASHERS

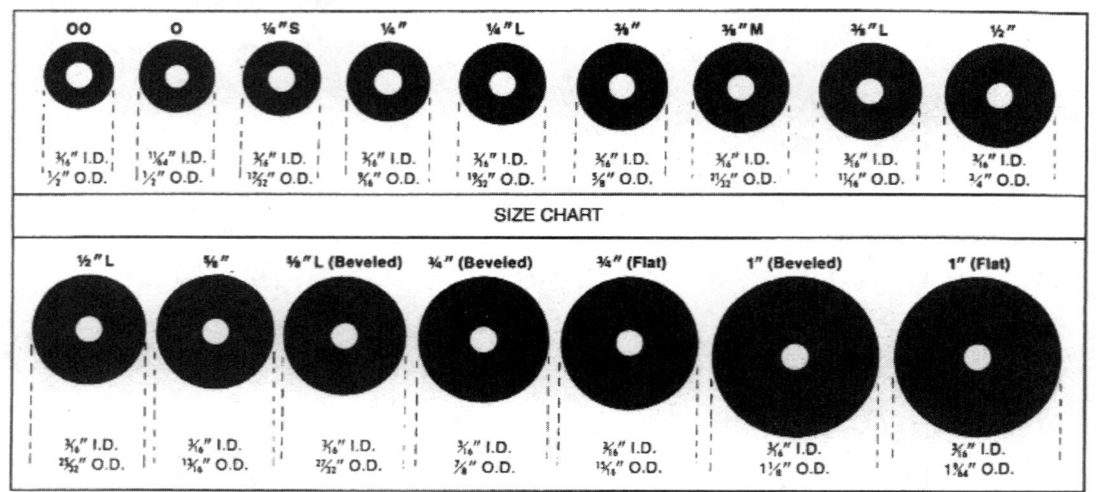

GOLDEN STATE® FAUCET WASHERS – HANDY BOXES							
W-230	00 Beveled	25	1 Box	W-245A	0 Flat	25	1 Box
W-231	1/4 S Beveled	25	1 Box	W-246	1/3 S Flat	25	1 Box
W-232	1/4 Beveled	25	1 Box	W-247	1/4 Flat	25	1 Box
W-233	1/4 L Beveled	25	1 Box	W-248	1/4 L Flat	25	1 Box
W-234	3/8 Beveled	25	1 Box	W-249	3/8 Flat	25	1 Box
W-235	3/8 M Beveled	25	1 Box	W-250	3/8 M Flat	25	1 Box
W-236	3/8 L Beveled	25	1 Box	W-251	3/8 L Flat	25	1 Box
W-238	1/2 Beveled	25	1 Box	W-252	1/2 Flat	25	1 Box
W-239	5/8 Beveled	25	1 Box	W-253	1/2 L Flat	25	1 Box
W-241	3/4 Beveled	25	1 Box	W-254	5/8 Flat	25	1 Box
W-245	00 Flat	25	1 Box	W-255	3/4 Flat	25	1 Box

Courtesy Radiator Specialty Co.

Index